GEOLOGY
FOR EVERYMAN

Sir Albert Charles Seward

GEOLOGY
FOR EVERYMAN

By the late

SIR ALBERT CHARLES SEWARD

With a Preface by

SIR HENRY LYONS

CAMBRIDGE

AT THE UNIVERSITY PRESS

1943

CAMBRIDGE UNIVERSITY PRESS
Cambridge, New York, Melbourne, Madrid, Cape Town,
Singapore, São Paulo, Delhi, Tokyo, Mexico City

Cambridge University Press
The Edinburgh Building, Cambridge CB2 8RU, UK

Published in the United States of America by
Cambridge University Press, New York

www.cambridge.org
Information on this title: www.cambridge.org/9780521238977

First published 1943
First paperback edition 2011

A catalogue record for this publication is available from the British Library

ISBN 978-0-521-23897-7 Paperback

NOTE

My husband had completed this book three days before his sudden death. He would, I know, have wished to thank many friends for their help, and this task now falls to me.

Dr J. J. Richey, of the Geological Survey of Scotland, and Dr J. Pringle were most helpful in reading the manuscript. Mr W. N. Edwards, of the Natural History Museum, not only read the manuscript, but completed the selection of illustrations. Professor C. E. Tilley was good enough to revise the book in proof. Mrs G. J. Kerrich (formerly my husband's secretary for many years at Cambridge) kindly undertook to prepare a typescript of the work.

My warmest thanks for the Preface are due to Sir Henry Lyons, one of my husband's oldest and most devoted friends; and, lastly, I have to thank the Syndics of the Press, with whose work my husband was long associated, for their readiness to undertake the publication of his last book.

MARY A. SEWARD

CONTENTS

v

PLATES

LIST OF FIGURES

PREFACE

My old friend, Albert Seward, Professor of Botany in the University of Cambridge from 1906 to 1936, was also a geologist of long and varied experience. He thus became an eminent authority on palaeobotany and wrote many learned works on the subject.

In his youth he had spent much of his leisure among the hills and river-valleys of Lancashire and the Lake District, collecting plant and rock specimens for study, and so formed the habit of careful observation to be developed by the numerous researches which he carried out in the course of a long and busy career. The lifelong interest which he found in these field studies suggested to him that there were probably many who would appreciate a clear and simply-worded account of the earth's surface and its phenomena, even though they might never have studied geology seriously. It was with this end in view that he wrote the present volume, which he had completed just before his death.

I talked over with him much of the subject-matter of the book. What Seward wanted to do was to emphasize the attraction which geology offers to anyone who enjoys a walk over the countryside, and his vivid memories of his own observations provided him with a rich harvest of material.

Assuming only a limited knowledge of geology in the reader, he begins with a simple summary describing the earth's crust and its constant changes. He then passes to the remains of animals, plants and other forms of life for many ages embedded in the rocks. After this preliminary survey the main purpose of the book is presented in the form of a series of journeys through the British Isles, so planned as to bring under review the geological characters of the various districts. In describing the deposits representing successive periods of the earth's history, Seward departs from the usual method and begins with the latest deposited rocks instead of with the earliest. Thus he describes first the conditions which prevailed when primitive man was giving place to the more developed human types of the later Stone Age, who cultivated crops and domesticated animals.

The long panorama of the earth's history, covering hundreds of millions of years, is revealed in descriptions of the terrestrial phenomena which characterized different periods—the Glacial period, for instance, during which for several hundred thousand years the greater part of these islands was in the grip of climatic conditions such as now prevail in Greenland. These extreme conditions gradually came to an end when intervals of milder climatic conditions made occupation by man and animals possible and characteristic vegetation invaded the country from southern Europe.

Moving back in the scale of time we come to the series of deposits formed when the climate was more temperate and the arctic conditions of the Glacial period were still in the distant future. While these tertiary rocks, as they are called, were being formed, plants and animals of more temperate and even sub-tropical regions appeared, and large areas, which were later to become land, were submerged beneath the sea. Wide regions were affected by these great movements of the earth's crust, massive deposits were laid down in the ocean areas, and, in later ages, mighty foldings of the crust produced the great mountain ranges of southern Europe and Asia.

This widespread submergence of large areas had been preceded by a period when much of the area described enjoyed a temperate and even tropical climate while the comparatively shallow seas contained a characteristic fauna; these are represented by the rocks in the belt of country lying to the west of the Chalk downs and other deposits of Cretaceous age. These, again, were preceded by a period when conditions were characteristic of an arid or semi-arid climate when salt lakes and desert regions resulted from the meagre rainfall.

By the combined evidence of the movements of portions of

the earth's crust and of the changes of climate which accompanied or followed them, a clear picture is formed of the changes which have taken place on the earth and the effects which they have had on animal and vegetable life.

In the north and west of England, in southern Scotland and extensively in Ireland, the history of these parts of the earth is to be read in the rocks of what are known as the Carboniferous series; a large series of deposits formed in the course of about a hundred million years as limestones in the open sea, and, later, the luxuriant forest vegetation providing the coal deposits, which have for years been of the greatest service to our industries.

The history of still earlier rocks has been preserved in Scotland, Wales and western England, where large fresh-water lakes once received sandy deposits carried into them by floods and rivers from a continent where a semi-arid climate prevailed; farther south an area of warm open sea favoured the formation of banks of coral, and a rich marine fauna.

Many of these earlier rocks have lost the traces of any fauna that they may have once contained by the pressure due to earth movements, and now form a folded and contorted series of beds which have been worn down to an irregular surface by the action of atmospheric conditions of rain, and snow, heat and cold, extending over a vast period of time.

To the geologist and also to those who have no technical knowledge, the evidence of past conditions presented in these chapters provides an inexhaustible fund of interest. It is an introduction which should encourage many to go more deeply into a subject which, apart from its bearings on many branches of industry, provides a knowledge of the earliest phases of primitive man and of the conditions under which he first inhabited parts of our earth.

H. G. LYONS

1943

CHAPTER I

Introductory: Geology as a Hobby

It is a commonplace saying, though not always a mere repetition of a familiar maxim, that we perform our ordinary duties more efficiently and with less mental strain if we have acquired the habit of switching our thoughts in leisure moments into channels other than those followed in the daily routine. On retirement many people with no hobby to spread gold-leaf on days to come exchange satisfying activities for boredom. The importance of cultivating one or more hobbies in early life is a subject frequently chosen by speakers whose search for topics appropriate to school speech-days is apt to be long and barren. A hobby, to be satisfying, must demand a certain amount of mental effort and should be such as can be progressively pursued until a stage is reached when a desire to know more opens up a prospect of work of one's own in a small patch of an unlimited field, modest in scope yet enthralling enough to produce the pleasant feeling of self-forgetfulness in research. It is not my intention to compare one hobby with another; my purpose is to draw attention to a branch of natural knowledge that is too often neglected and not infrequently avoided because of disinclination to learn a few unfamiliar technical terms. The science of geology, or earth-history, like any other science, has its own terminology, to which fresh names are constantly being added to fit new discoveries: the difficulty of technical terms is rather imaginary than real; Shelley was disposed to think

> human pride
> Is skilful to invent most serious names
> To hide its ignorance.

It may, however, be suggested that some of the technical terms in common use by geologists should find a place in the vocabulary of all educated people. A knowledge of geology sufficient to illustrate the possibilities and the fascination of the subject can easily be acquired without the trouble of learning a new language. All that is needed is ability to observe what is happening before our eyes and to apply knowledge gained from the present to the interpretation and reconstruction of the past. Geology is not one subject but many, and may be described as the application of all sciences to a comprehensive and co-operative investigation of the earth on which we live. The aims of a geologist are wide and ambitious, he inevitably tends to become a specialist and concentrates attention on a few branches of the subject. On the other hand, if we have within us the spirit of adventure and the passion of the search, we can all learn the art of deciphering some of the pages of Nature's open book. Geology is unfortunately rarely taught in schools and young people have little chance of discovering whether or not it is likely to appeal to them. Arguments in favour of its introduction into school curricula were put forward in a British Association Report on scientific education in 1867 and again in two Reports in 1936 and 1937, but the recommendations met with only partial success. It is obviously difficult to make additions to curricula already over full, and yet if a member of the teaching staff of a school happened to be interested in the subject he could encourage field-work out of school hours and co-operate with the pupils in cultivating an ideal hobby.

Rocks and fossils are not merely things to collect; they are scraps from Nature's story-book which give us glimpses of stirring events, revolutions in the physical world alternating with periods of comparative stability, the unfolding of life as age succeeds age. Slight acquaintance with the methods of geologists and a little knowledge of the subject enable us to see the world through magic spectacles, to look beneath the surface into the depths below and backward from the present to a past inconceivably remote.

My chief aim is to present a case for the inclusion of an intelligent interest in geology as part of that intellectual equipment we call general culture, culture that has been defined as what remains after we have forgotten all we learnt at school. Whatever our business in life may be, we can read in the rocks and the fossils they contain some at least of the blurred and fragmentary pages of a chronicle of the pageant of Nature far transcending human history both in the period it covers and in its appeal to the imagination. Among the more vivid recollections of my boyhood are days spent alone looking for fossils on the northern shores of Morecambe Bay; it mattered little that my knowledge of the things found was extremely meagre; the discovery of pebbles on the beach excited my curiosity and a desire to know more about the vanished world of which they had been a part. Having collected specimens well enough preserved to be easily recognizable as the remains of creatures that were once alive, it is natural to go a step farther: by consulting books and looking at fossils in a museum we can supply names and obtain information on the relationship of shells and corals to living forms. Thus the pleasure of the search is increased and we begin to understand the significance of the things found.

The descriptions given in the following chapters are not intended to serve as an elementary text-book; my hope is that they may be used as stepping-stones to something higher and more scientific. A geologist is no less sensitive than other people to the beauty of a landscape; he sees in surface features the work of Nature's sculpturing tools: from rocks exposed in cliffs and ravines he is able to follow successive stages in the development of the present from the past. To his enjoyment of the beauty of a scene is added a deeper sensation that comes from closer contact with the mysteries of Nature and the epic of creation. After enjoying to the full a view suddenly revealed as the mountain top is reached, a geologist with mind prepared by what he has learnt of earth's story can hardly repress a desire to look backward to earlier scenes. He sees in imagination the same part of the land that is now called England as it was before the hills had been uplifted; the real land becomes a land of dreams; time is put back millions of years. Hills and valleys give place to a glistening expanse of blue sea. As a less remote past comes into view dark domes of rock heave into sight above the face of the water and by degrees the new-born land grows higher; sea is replaced by ranges of hills that in the course of ages rise to heights far exceeding those of the actual present. Torrents of turbid water scoop out deep gullies in the mountain flanks; avalanches tear away rocks, and showers of stones riven from crags by recurring frosts co-operate in the wholesale destruction. 'Every valley shall be exalted, and every mountain and hill shall be made low', 'He turned the sea into dry land' are more than poetic imagery. Mountains of to-day are relics of lofty alps which in a previous age rose from the ocean-bed where their foundations were laid by long-continued outpourings of submarine lava and showers of volcanic ash or by the accumulation of gravel, sand, and mud carried by rivers from a neighbouring continent. A slight knowledge of earth-history adds enormously to the enjoyment of scenery: the contours of hills—smooth slopes and jagged peaks, wild moorland and fertile plain—acquire a deeper meaning when we think of them as the expression of natural forces acting upon various kinds of rock of which the surface of the land is made. All of us have memories of scenes and places which live in the light of the morning star of memory. One such memory of early days is Easedale Tarn within a short climb from Grasmere, an undistinguished yet impressive and solemn patch of water in a rocky cup below the Palaeozoic cliffs of Lakeland, which seems to set back the clock, carrying our thoughts far across the ages to scenes made visible by our imaginings, stimulated by turning over the pages of Nature's manuscripts. Among the plants covered by the water near the edge of the tarn grows the quillwort (*Isoetes*), an unfamiliar and little regarded plant with a tuft of green spiky leaves borne on a small stumpy stem, a distant relation of the more familiar club-moss (*Lycopodium*): this plant of mountain tarns is in harmony with the spirit of its environment; it is a symbol of continuity, one of the oldest relics in the vegetable kingdom, a diminutive descendant of greater and more complex forbears which lived in forests of an age separated from the present by at least 300 million years.

The ancient Egyptians pictured the earth as the floor of the universe. In the Middle Ages men began to think of the world as older than was commonly believed and laid themselves open to the charge of heresy. For a long time, and largely as a concession to the literal acceptance of the Hebrew story of Creation, Noah's flood was quoted as a satisfying explanation of the occurrence in unexpected places of animals and plants embedded in rock. Belief in a universal deluge persisted until the middle of the nineteenth century; but as always there were a few men who had a clearer vision, and among them that universal genius, Leonardo da Vinci (1452–1519), who saw in fossil shells preserved in rocks far above sea-level proof of oscillations of land and sea. The presence of sea-shells in rocks hundreds of feet above the present level of the sea has often been a cause of popular misconception. One hears people say, when told that chalk, for example, was formed on the floor of an ocean, 'Then there must have been water over the downland'. They think of the hills as everlasting and fixed, and fail to realize that we live on an unstable and mobile earth; the rock that we call chalk was raised from the bed of the Cretaceous sea in the course of one of many foldings and upliftings of the earth's crust. A great advance was made rather more than a century ago when Sir Charles Lyell published a book entitled *Principles of Geology* in which he developed the thesis that the present is the key to the past. He showed that there is no need to have recourse to universal floods, catastrophes which wrecked the earth's surface, and other miraculous events; his contention was that the forces now moulding the earth's face—water, frost, and wind, volcanic activity, and earthquakes—have always been operative as they are to-day. Their effects to our restricted vision seem hopelessly inadequate, but when millions of years are substituted for centuries we are able to appreciate their potency. Continuity and not sudden and devastating cataclysms is the doctrine preached by Lyell. Through the long course of geological history there have been zones of climate, tidal movements of water, circulation of the air, sunshine and rain; there has been little change in the nature and working of the world's machinery; on the other hand, there has undoubtedly been a wide range in the intensity of physical forces and in the evolution of the living world.

There is no clearly defined line between geology and geography: a geographer is not only interested in the surface of the earth as he sees it; he is also interested in the origin of physical features and the part played by air and water in fashioning the rocks into their present shapes. He leaves to the geologist the history of plants and animals recorded in the rocks. The geologist endeavours to envisage the distribution of land and sea and to construct geographical maps representing different stages in the evolution of the world. He, unlike the geographer, is not primarily interested in the earth's surface as a stage on which men play their parts; his aim is to follow the kaleidoscopic pattern imperfectly imprinted upon the rocks which are the sources available to the historian of vanished lands and seas.

The part played by man is an example of another difference between geologists and geographers. The human race is one of the more recent products of evolution; though we cannot assign a date to the birth of man, we know that he is but a creature of yesterday in comparison with the whole span of geological history. For the most part geological history is confined to ages preceding by hundreds of millions of years the earliest traces of animals worthy to be called ancestors of creatures made in the likeness of man. A world without man connotes more than we can imagine without a supreme effort: we inevitably think of the world in relation to man's welfare, as his playground, his workshop, his battlefield. We are apt to forget that the heritage he enjoys was not made for him alone; beauty is eternal and an attribute of Nature old as creation. The beauty of Nature in the past can be recaptured only in imagination. It is certain that the sun, moon, and stars, wind and cloud, rain and storm, the music of the sea, rainbow-coloured mists, colour in the dead as in the living world, have persisted through the ages. There are few people who are not consciously or unconsciously influenced by natural beauty: the sun slowly sinking behind the edge of the sea; the ethereal delicacy of the greenish blue sky streaked with bands of a red brilliance; a panorama of mountain peaks; the glory of woods in autumn. Each of us can recall scenes on land and on sea that have stirred our hearts, raising us to a higher plane, setting in motion vibrations which seem to be an expression of a force that is not of this world, a mysterious influence bringing us nearer to the eternal values. We experience an exaltation, a sense of being permitted to have a glimpse of the masterpieces of creation. It is this sense of approach to the infinite that is awakened when we are able to follow the working of a creative force that has operated ever since the world began. The privilege of being able to make contact with the past gives us something more than mere satisfying of curiosity; it gives us a keen perception of the Creator's methods. As we interpret the hieroglyphs written by Nature we gain not only intellectual but also spiritual refreshment. The more we know of the history of the earth the more we marvel, the more deeply we feel the inadequacy of a conception of the universe as the soulless product of physical and chemical processes fortuitous and uncontrolled.

Many people think of the face of the earth as a face in repose and, if freed from man's interference, almost changeless. The geologist sees in the rocks of which the earth's surface is made marks of ancestral characters which speak to him of changing landscapes and of the march of events through countless ages. Before setting out on our journey along the corridors of time we shall do well to accustom ourselves to the habit of thinking in terms not only of thousands but of millions of years; we must get away from the hampering influence of the time-standards of human history. Features of the earth's surface familiar to us from childhood seem to be emblems of permanence. We see in them examples of Nature's enduring monuments in contrast to the rapidly shifting scenes in human history. It is essential to remember that the infinitely small becomes amazingly large when time is reckoned on the generous scale permitted to geologists. Destructive and constructive operations act slowly and gradually, and on a superficial view results are scarcely noticeable. If, on the other hand, a measure is taken of the almost imperceptible effects of rain and frost, and if we think in terms of thousands of years, the enormous power of these agents becomes apparent.

In his most fascinating book, *The Mirror of the Sea*, Conrad lets us share with him thoughts on the age of the earth as he contemplated the ocean during a watch on the bridge of a ship: 'If you would know the age of the earth, look upon the sea in a storm. The greyness of the whole immense surface, the wind furrows upon the faces of the waves, the great mass of foam, tossed about and waving, like matted white locks, give the sea in a gale an appearance of hoary age, lustreless, dull, without gleams, as though it had been created before night itself.' One may also draw a comparison between the ocean lashed by a gale—a never-to-be-forgotten sight—and rocks under the influence of titanic forces operating through the earth's crust during one of the recurrent storms that left their impress on the rocks. Waves in a moderate breeze form series of regular, rounded arches and troughs: similarly layers of rock have been pressed by lateral forces into regular corrugations. Under the propelling might of a gale the arched waves are impelled forward until their backs, exposed to the storm, curve in great curving folds and downward sweeps, the arches are overfolded. As we shall see later, in intensely folded regions rocks behaved as though they were fluid as water, arched folds were bent over until they became overfolds and recumbent folds.

Anyone, other than a geologist, to whom a fossil is shown, usually asks—how old is it? Many attempts have been made to supply an answer to this very natural enquiry; some were based on data obtained by measuring the rate of processes now in operation. It has been said that our island is being worn away at the rate of one foot in 3000 years, an estimate which enables us to make a guess how long it would take to reduce the whole country approximately to sea-level. But we know that in the course of ages the levelling of mountains, which has repeatedly occurred, is followed by fresh upheavals and the building of new ranges of hills. Geological time cannot be measured with precision by methods such as this. Another method of attacking the problem was adopted by Lord Kelvin rather more than seventy years ago. The earth was once part of the sun and when it began its existence as a separate entity, at first gaseous and molten and later cool enough for rocks to crystallize on the surface, an internal store of heat remained as a legacy from its torrid parent. This store would gradually and steadily diminish as heat escaped by radiation into space. Starting from this assumption, Lord Kelvin calculated how long it would take for a molten ball the size of the earth to lose an amount of heat represented by the difference between the original and the present temperature. The conclusion reached was that the earth's age is 100 million years; this he afterwards reduced to a much more modest estimate. There was no reason to suppose, and, indeed, it would have been contrary to experience to suppose, that any fresh accession of heat could be obtained from the earth's substance. The discoveries of one age often invalidate those of a previous age: a striking illustration of this is furnished by the discovery of X-rays and radioactivity, which, as we shall see, demonstrated the fallacy of Lord Kelvin's methods. As Professor Joly of Dublin said: 'We are dwellers upon a world in the surface material of which there exists an all but inexhaustible source of heat.' In 1895 Professor Röntgen of Würzburg was experimenting in his laboratory on electric sparks passing through a tube from which the air had been removed: wishing to cut off light from the sparks he covered the tube with opaque material, but he noticed that some kind of radiation still came from the tube and affected photographic plates in the neighbourhood. This was the first important step towards the discovery of X-rays. A year later Professor Becquerel of Paris found that rays like those detected by Röntgen were given off by the element uranium. This was the beginning of the science of radioactivity. In 1898 Madame Curie isolated from pitchblende a new element, radium. Uranium and another heavy element, thorium, which occur in minute quantity in many rocks, possess the extraordinary property of continually breaking down into other elements, one of which is radium. The name 'element' was formerly given to different kinds of matter which it was held could not be broken up into anything unlike themselves: uranium and thorium are exceptions to this rule. Radioactive substances, which are traceable to the two elements uranium and thorium, constantly disintegrate; they break down into other substances and their disintegration is accompanied by liberation of energy in the form of heat. Radium is continuously giving out heat, and the source of most of this liberated energy can be traced to the high velocity of expelled atoms of a gas, helium. The element into which radium is transformed by loss of a helium atom and part of its latent store of energy, undergoes further disintegration and eventually a stable end-product is formed which is a special kind of lead, known as uranium-lead; being stable it accumulates in rocks where it was formed. Lord Rayleigh found that radioactive substances are widely distributed in rocks. By calculating the amount of uranium and uranium-lead in samples of rocks of different ages he obtained data from which it was possible to estimate the age of a rock in years. The rate of production of lead from uranium is known: 1 million grammes of uranium give rise to 1/7500 of a gramme of lead every year, and it is important to note that the behaviour of radioactive substances is not affected by physical or chemical conditions. Hence Professor Holmes, in his book *The Age of the Earth*, was able to write: 'Certain radioactive minerals—more precious than gold—are literally natural chronometers registering time within themselves and revealing the passage of hundreds of millions of years since the time they first took form as part of the rocks and mineral veins in which they occur.' The age of the oldest known rocks is at least 2000 million years, and it is reasonable to assume a still greater antiquity for the actual foundation-stones of the earth's crust. The present method of measuring geological time has multiplied Lord Kelvin's estimate more than a hundredfold. It is now possible not only to refer rocks to their relative age, that is, to assign them to a particular geological period, but to speak of their age in terms of years.

Geology covers a wide field embracing the rocks and fossils of the whole world: geologists endeavour to unravel the life-history of the earth since it ceased to be gaseous and molten and to follow the unfolding of life from one epoch to another; they take over the youthful earth from the astronomers, and

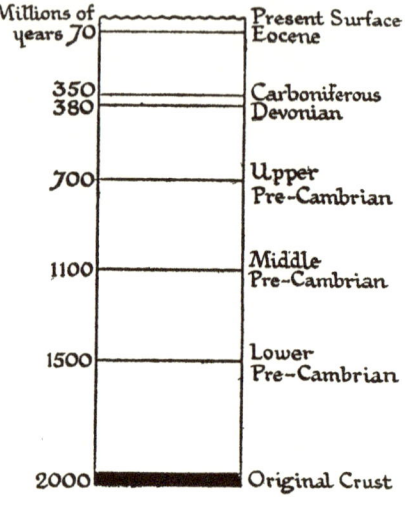

Diagram of the age of the earth

in deciphering the later stages of earth-history they seek the assistance of archaeologists who in turn give place to students of written history. How, it may be asked, is it possible in a few short chapters to convey an adequate idea of the science of geology? All that is attempted is an introduction to methods of geological enquiry illustrated by descriptions of a few scenes from the past reconstructed from rocks and fossils. We who live in the present can watch the building of rocks in seas and on land; we can often discover whence came the material of which rocks are made; we can study living animals and plants. It needs but a little geological knowledge to live also in the past and to find there a new meaning and an enormously enhanced enjoyment of the present. Thomas Carlyle in *Sartor Resartus* cynically remarked: 'It has come about that now to many a Royal Society the creation of the world is little more mysterious than the cooling of a dumpling.' In the early stages of knowledge problems seem to be relatively simple; but as wisdom follows, the mystery deepens. Contributions to geological science since Carlyle's day have not solved the problem of creation; the corner of the veil hiding the inner courts of knowledge has been lifted ever so slightly. As we travel a little farther along the path of adventure we are confronted with still more problems, many of which will never be solved. None the less, with such knowledge as is within our reach,

> We look at all things as they are
> But through a kind of glory.

As more of Nature's secrets are revealed 'Eternity and infinity lose part of their strangeness'.

CHAPTER II

Earth's Story-book

A student of human history endeavours to put himself into the position of a spectator of events that happened centuries ago; in order to see them as though they were taking place before his eyes he consults contemporary records, biographies and other sources from which to bring to life the actors and envisage the environment. Historians whose aim is to describe conditions of life and the habits of people in days when man had not reached the threshold of civilization search for evidence of another kind; they collect flint implements, bronze weapons and ornaments and, when possible, human skeletons. Light may be thrown upon conditions of life by the discovery of old settlements and primitive dwellings where shells and bones in refuse heaps indicate the nature of the food supply and of the animals contemporary with man. The earlier stages of human history known as pre-history are reconstructed not only from objects found in places where they were left by man, but in places where they were strewn all too prodigally by Nature's hand. Here then we have a period transitional between human and geological history. The student of earth-history makes use of sources provided by Nature; like the authors of ancient, mediaeval, and modern histories, he finds it convenient to adopt distinguishing names for periods characterized by outstanding events or marking phases in development. The sources from which geological history is compiled are necessarily very incomplete and often fragmentary and yet they give us glimpses of far-off days in a world that knew not man.

It is customary to speak of the solid framework of the world within reach of investigation as the earth's crust, a convenient name suggested by the analogy of the solid crust which forms over a cooling molten mass. The crust of the earth consists not only of rocks visible at the surface but also of rocks hidden from view. As will be shown later, it is often possible by noting the relative positions of different layers of rock and their angles of inclination to form a general idea of their continuation downwards. Borings also tell us much of the composition of the crust at depths of a mile or more. In general terms the crust includes the rocks of which continents are built as well as those underlying the ocean-floors. What do we know of the inaccessible and deeper parts of the earth? It was formerly believed that the interior is molten, a view apparently supported by the fact that lava, which is molten rock, is poured out over the surface during volcanic eruptions. Many people now believe that the earth is solid to the core; another recent view is that the core is liquid. A solid earth would seem to be inconsistent with recurrent outpourings of lava, but it does not necessarily follow that lava comes from a subterranean source where the rocks are always molten. When a solid passes into the liquid state its particles are forced farther apart under the expanding influence of heat; they move more freely; if pressure is applied to the surface of a solid that is not free to move laterally and is forcibly confined on all sides a much higher temperature is required to alter the position of the particles and so melt it. The greater the weight of the load the greater will be the amount of heat necessary to liquefy the rock and convert a solid into a liquid or a plastic mass. Rocks deep below the surface may be solid even though the temperature is far above the melting-point of the same rocks under a lower pressure. Forces of compression and tension are constantly acting upon the earth's crust; rocks are exposed to strains which from time to time become too great to be withstood and the solid ground cracks and splits with the formation of rifts and fissures. This obviously relieves the pressure; rocks that were solid become plastic and melt, and as the water imprisoned within them flashes into steam the semi-liquid material is forced upwards towards the surface and some flows over the ground as lava. Drops of liquid imprisoned in cavities in a crystal are relics of condensed vapour within the originally molten mass. In several parts of the world many thousands of square miles are covered by sheets of lava which at various times were forced upwards from subterranean reservoirs and spread as devastating floods over the surface. These rocks of deep-seated origin are relatively dense: there is reason to believe that more deeply seated rocks are denser still. The earth as a whole weighs more than five and a half times a globe of water of equal size: the crust is only two and a half times as heavy as an equal bulk of water. It is therefore obvious that the interior must greatly exceed in density the outer layers. Meteorites, the balls of fire mentioned in the Psalms, which often consist mainly of nickel and iron, are samples of the composition of extra-terrestrial bodies and may afford a clue to the nature of the heavy material at or near the earth's core. Another clue is furnished by the rate of travel of earthquake waves, which is conditioned by the nature of the rocks traversed. We cannot probe far below the surface, even in the deepest bore, but as was said by one of the earlier writers on earth structure more than a century ago: 'Men can see much farther into the interior of the globe than they are aware of'; he goes on to say that geologists are blamed without reason for theorizing on the earth's interior when all they can do is to make a few scratches on the surface.

Among the most spectacular results of geological research is the certain knowledge that the distribution of land-areas and oceans has varied within wide limits in the course of ages. One of the many aims of geologists is to draw maps of the world

as it was at different periods, most of which bear little resemblance to those with which we are familiar in modern atlases.
The data on which such maps are constructed are necessarily
too incomplete to provide information for more than tentative
reconstructions which reflect differing opinions of authors.
Rocks containing marine fossils are guides to the plotting of
former seas; others that furnish evidence of shallow-water and
shore conditions or of actual land-surfaces are aids to the position of continents. Lands now far apart and separated by comparatively deep seas are incorporated into a single continent.
Assuming that there is valid evidence in support of taking
liberties with the arrangement of land and water, a natural
question to ask is—how can the changing geographies be
explained? Let us take two examples: it is generally agreed
that Greenland was once united to western Europe and that
South America and the southern half of Africa were either
joined together or at least lay very close to one another. There
are two possibilities: one is that these land-masses were formerly
united by a connecting bridge, either broad or narrow. This old
land-connexion foundered and sank below the sea, a supposition in some instances supported by ocean soundings revealing
submerged ridges. But many examples might be quoted of
land-bridges postulated by geologists which have left no trace;
all that can be said by those who believe in them is simply that
they were, and now are not; they foundered. Belief in the
former union of continents now sundered by oceans is based
partly on the striking similarity of rocks and fossils in the two
widely separated regions, partly also on many facts furnished
by a study of the geographical distribution of living animals
and plants. Certain land animals, either specifically identical or
closely related, are known to occur in continents now too far
apart to admit of migration or transport across a barrier of sea.
Similarly, even admitting that fruits and seeds can be carried

long distances by wind, by birds, or by floating wood without losing their vitality, such means of transport are in many instances inadequate as an explanation of the occurrence of the same kinds of trees and herbs in regions separated by broad oceans. The subject of geographical distribution of animals and plants, like many others not strictly geological, cannot be ignored by students of earth-history as it is their business to make use of all sources of natural knowledge which contribute to a better understanding of the present in relation to the past. Darwin spoke of that 'noble subject' geographical distribution, which plays a prominent part in the presentation of the case for evolution in the *Origin of Species*: its importance is clear if we accept Darwin's view expressed in the following passage: 'It is also obvious that the individuals of the same species, though now inhabiting distant and isolated regions, must have proceeded from one spot where their parents were first produced.' He goes on to say: 'The simplicity of the view that each species was first produced within a single region captivates the mind. He who rejects it, rejects the *vera causa* of ordinary generation with subsequent migration, and calls in the agency of a miracle.' The subject of geographical distribution is mentioned here simply because it raises many problems which are insoluble unless we can employ information gained from the rocks, thus helping us to visualize the relation of land to sea in former ages in its bearing upon the all-important question of migration routes along which dispersal of animals and plants may have been effected. The past is the key to the present. Former land-connexions have been a fruitful source of controversy and acute differences of opinion; the main point is that there must have been either actual land-bridges or at least chains of islands as stepping-stones between continents now separated by impassable barriers of water.

The second possibility—an alternative to the disappearance

of former connecting bridges—is one which was revived in a new form and with new arguments not many years ago by the late Professor Wegener, whose tragic death in Greenland deprived science of an able and courageous investigator. Wegener's hypothesis, as it is usually called, provides another and very different explanation of widespread changes in the relative positions of land and sea which appear to be demanded by evidence furnished by the rocks and by the distribution, both now and in the past, of animals and plants. Briefly stated, this second alternative is based on the supposition that the continental areas of the world have shifted their positions by moving laterally; there has been, so Wegener believed, drifting of continents just as icebergs are passively carried by ocean currents. Oceans and continents are regarded as fundamentally distinct; that is to say, the floor of the oceans is not simply a continuation downwards of the continental rocks of the land, but consists of a different layer heavier than that of the land-areas. Wegener estimated the thickness of the rocks of which continents are built at approximately 60 miles, and he described this layer as floating on a lower layer of the earth's crust from which the upper and lighter layer projects to a height of rather more than 3 miles. He also believed that the heavier layer extends under the ocean-floors. The relatively light rocks composing the continents do not form a layer of uniform thickness; in some places, e.g. where high mountain ranges occur, the upper and lighter part of the crust is thicker and sinks lower into the heavier layer below. It is generally agreed that up to this point the hypothesis expresses the true state of affairs, namely the twofold nature of the crust, an upper, lighter layer made of the rocks we see on the earth's surface and a lower, heavier layer consisting of rocks known as basalt which are practically identical with the majority of lavas erupted from volcanoes now and in former ages. It is this basaltic layer on which the oceans rest. Basalt is the rock of which the Giants' Causeway in northern Ireland is made; it also plays a large part in the construction of many of the Inner Hebrides. We come now to a more controversial aspect of the Wegener hypothesis.

A glance at a map of the Southern Hemisphere shows a fairly close correspondence in outline between the coast of Brazil and the west coast of Africa on the opposite side of the South Atlantic Ocean. Similar correspondences have been observed in other countries that are now far apart. At the present time the continents, with their submerged margins projecting as shelves covered by shallow water between the shore-lines and the ocean deeps, occupy only one-third of the earth's surface: Wegener supposed that originally the material of which the continents consist was spread as a thin film over a large continuous area of the earth's surface. As this film contracted in the course of millions of years great rifts were formed in it and as they slowly increased in breadth blocks of the ruptured surface-rocks drifted farther apart as slabs floating in the heavier and semi-plastic basalt that lay beneath. The upper portion of the crust was not only broken up into separate blocks but the detached areas were reduced in superficial area by compression and folding of the rocks. Wegener envisaged the primaeval land as a board before it is cut into pieces of a jig-saw puzzle. When the severed pieces are moved apart along a flat surface they are treated as he supposed the slabs of a dismembered continent were treated by natural forces (Fig. 1). One of the weak points in Wegener's hypothesis is that no satisfactory answer has so far been given to the question—what was the nature and the manner of operation of the forces? Wegener cited as an argument in favour of drifting continents discrepancies between observations on longitude made at certain localities in Greenland on three occasions. In 1823 measurements were made on Sabine Island off the east coast of northern Greenland; in 1870

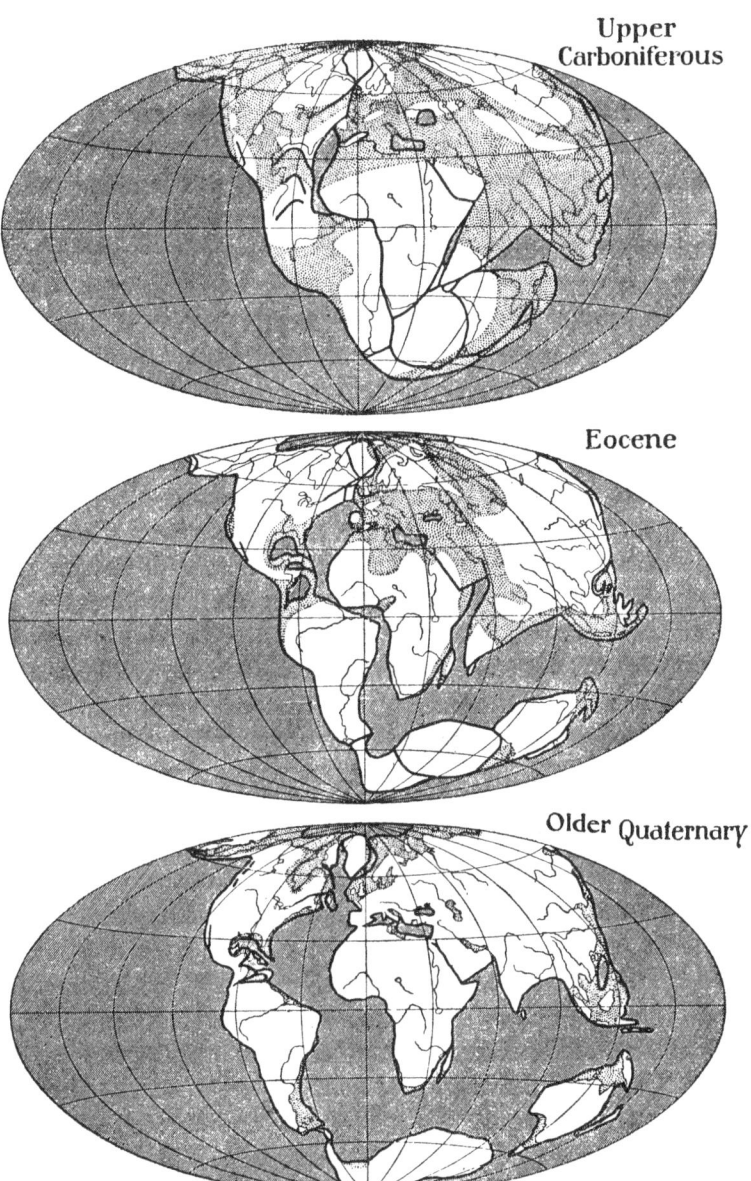

Upper Carboniferous

Eocene

Older Quaternary

Fig. 1. Reconstructions of the map of the world for three periods according to Wegener's hypothesis.

at a place about 100 yards farther east; and in 1907 a third series at a place farther north. Comparison of the results seemed to show that the distance of Greenland from Europe had progressively increased at a rate of approximately 9 metres a year between 1823 and 1870 and 32 metres a year between 1870 and 1907. There is, however, no doubt that these estimates are untrustworthy because the measurements were not made with a precision that is possible only when refined methods are employed; moreover, they were not made at the same locality. In 1927 and 1936 measurements were made under much more scientific conditions by members of the Danish Geodetic Institute with a modern transit instrument and from the same pillar on the west coast of Greenland. The results were identical. It must, therefore, be admitted that as yet proof in support of drifting continents is lacking. On the other hand, Wegener's hypothesis, or some modification of it, need not be entirely abandoned. It is possible that in the future decisive evidence may be forthcoming. The conception of drifting continents is attractive and tempts some of us to indulge in wishful thinking because it would help towards the solution of many problems such as the remarkable correspondence between continents now far apart and certain very baffling facts of geographical distribution of animals and plants both at the present day and in times past. My excuse for mentioning a subject much too difficult and controversial to be discussed more fully in a book intended for the general reader is that it may serve as an example of a speculative enquiry into a branch of geological history distantly related to geology as the science is usually understood, and it may also awaken an interest in a theoretical question which has been frequently debated in recent years.

One of the pioneers of the modern science of geology said that geological evidence affords 'no trace of a beginning, no prospect of an end'. Astronomers tell us that the earth began its separate existence as a gaseous, molten globe detached from the parent sun; in the course of its early life-history as an independent entity the surface solidified and the youthful earth entered upon the first phase of its geological career at a time separated from the present by 1500 or 2000 million years. Is it possible to recognize in the rocks relics of the original crust, to answer the question—'Whereupon were the foundations thereof fastened'? It has not so far been possible to discover the primaeval foundation-stones: the first products of crystallization of the molten material probably sank beneath the actual surface and it was not until later that a crust was produced in permanent form. Further reference is made to the oldest known rocks in a later chapter.

It is obvious that the rocks accessible to us cannot all be of the same age; ever since the earth's crust was formed there have been in ceaseless operation two processes: the wearing away of the surface by rain, frost, and other eroding agents, and concurrently, the piling up of the products of destruction—sand, mud, and the like—as well as the accumulation in seas, lakes, and oceans of myriads of shells and other remains of animals and plants. There has been rock-destruction and rock-building through the ages. The crust of the earth is made of layers of rocks of many kinds and if there had been no movement and dislocation of the surface, throwing rocks into confusion, the layers would form a regular series like so many sheets of paper or courses of masonry; it would be a simple matter to draw up a table showing their relative ages. But this is far from reality. It has, however, been possible with the help of fossils and various other guides to assign the component parts of the crust to different periods of geological history, the age and duration of which can now be estimated with a closer approximation to the truth than was formerly possible. The rocks of each period are distinguished by special names, some denoting districts in the British Isles or on the Continent, where knowledge of certain series of rocks was first gained, such as Cambrian from Cambria, the old name of Wales; Ordovician from the home of the Ordovices, a British tribe; Silurian from Silures, another British tribe on the English-Welsh border; Jurassic from the Jura Mountains on the western border of Switzerland. Other names were adopted because of the occurrence of well-defined and easily recognizable rocks such as Cretaceous from the Latin word for chalk, *creta*, Carboniferous from coal, which is mainly carbon. These and other names should be committed to memory, and their relative position in the geological table (p. 26), in order that descriptions of the periods in the following chapters may be more easily followed. Geological history is conveniently divided into major subdivisions called eras, and smaller subdivisions known as periods; the latter are again subdivided but it is not necessary for present purposes to remember the names. The names of the major groups or eras—Palaeozoic, Mesozoic, and Cainozoic (often called Tertiary)—are from Greek words meaning ancient, middle, and new or recent life and have reference to stages in the evolution of animals and plants from the earliest period at which recognizable fossils occur in the rocks. In the oldest rocks no fossils have been found: there was a time when the world was without life but at what stage in the history of the earth the first germ of life appeared we shall never know. In certain parts of the British Isles, especially in the North-West Highlands of Scotland and in many regions throughout the world, there are thousands of feet of rocks of different kinds in which no satisfactory traces of animals and plants have been discovered; these rocks are most conveniently grouped together as belonging to the pre-Cambrian era, and the oldest of them are often spoken of as Archaean. The term pre-Cambrian simply means that all the rocks so named are older than the lowest and oldest members of the Palaeozoic era. As we shall see later, despite the fact that pre-Cambrian rocks have so far yielded only problematical fossils, there can be no doubt that there was life on the earth long before the end of that era. It is certain that the duration in time of the pre-Cambrian era was far longer than all succeeding eras put together and must be reckoned in many hundreds of millions of years.

The grouping of rocks according to their relative age and on a scientific basis was a most important step in the progress of geology: credit for this must be given to an Englishman, William Smith (1769–1839), son of a blacksmith and assistant to a surveyor at Stow-on-the-Wold, Gloucestershire. Discoveries made in the course of his duties as an engineer and surveyor in the country near Bath earned for him the title 'the Father of English Geology'. He noticed that the rocks which he examined contained many fossil shells and that different series of rocks were characterized by different suites of fossils. This led him to the important conclusion that rocks occur in a regular and constant order of superposition and can be recognized by the fossils they contain. He subsequently extended his observations to other parts of England and found that deductions made from facts noted in one area applied equally to other areas. In 1815 Smith published the first geological map of England and Wales.

The history of the earth can never be written as a complete and continuous narrative: the sources are lamentably incomplete; in the course of recurring disturbances in the crust rocks have been made and unmade, and the records of life have been partially or completely destroyed. Our first task is to classify Nature's tablets as we arrange written records in the order of their antiquity before attempting to compile from them as connected a story as possible. The two principal criteria employed in fixing the age of rocks are (i) the order of their arrangement as strata or layers of the crust, (ii) the nature of the fossils they contain. The first of these is not invariably a trustworthy guide: though it is usually safe to assume that if we find series of beds lying one over the other the lowest is the oldest and the uppermost the youngest, there are some notable exceptions. A remarkable example of inversion of the original order of superposition is described in a later chapter in the account of rocks in the North-West Highlands of Scotland. There are, moreover, many rocks that are wholly or in part barren, having no recognizable fossils and therefore giving no clue to the nature of the animals and plants in existence when they were formed; but these and other difficulties are seldom

GEOLOGICAL TABLE

Eras	*Periods*
CAINOZOIC	Quaternary (including Pleistocene) Tertiary
MESOZOIC	Cretaceous Jurassic Triassic [1]
PALAEOZOIC	Permian [2] Carboniferous Devonian (including Old Red Sandstone) Silurian Ordovician Cambrian
PRE-CAMBRIAN	The older rocks of this era are often spoken of as Archaean

[1] This name was first used in Germany where the rocks of the period were divided into three groups, a Triad. In the British Isles only two of the groups are represented.

[2] From the province Perm in Russia, where rocks of this period were first distinguished.

insuperable obstacles in the way of drawing up a table of contents of the history-book of the earth. In the preceding table are included eras and periods only; the smaller subdivisions, a few of which are mentioned later, are omitted. Each era may be regarded as a separate volume and the periods as groups of chapters comparable to dynasties or lines of reigning sovereigns in human histories. The boundaries drawn between eras and periods are in the main arbitrary: apparent discontinuity is due to the imperfection of the record and to our ignorance. Darwin, as a young man face to face with Nature in Tierra del Fuego, was impressed by what seemed to him to be evidence of discontinuity between one set of rocks and another. He wrote in his Diary: 'A geologist perhaps would suggest that the periods of Creation have been distinct, remote the one from the other; that the Creator rested in his labours.'

CHAPTER III

Destruction and Reconstruction

Rocks crumble into dust and from the dust new rocks are made; matter is indestructible; change and decay are stages in constructive processes. What is called destruction is not annihilation, it is an undoing of something which was put together, the conversion of a complex structure into its component parts. We wish to know what rocks are made of, how they were made and how we can interpret them as records of earth-history. It is fitting therefore that we should first learn what we can by looking at the products of their disintegration and by so doing form an idea of the methods of rock-construction at the present day. What is happening now happened in the past. Since the earth became solid, destruction and reconstruction have been unceasing: on the stage of Nature the scenery rapidly changes when we watch the procession of events foreshortened in the tract of time: the forces controlling them—the actors in the drama—remain the same.

The sea encroaches on the land in some districts; in others we see ancient seaports converted into inland towns by the growth of new territory. The coast-line of England seen by Julius Caesar was not precisely what it is now. It is highly probable that our earliest ancestors needed no boats to take them from Britain to Gaul: it is certain that Neolithic man roamed through forests and across gloomy fens which stretched far beyond the present East Anglian coast. The greater changes in the coast-line are due not only to the wearing away of land by sea, they are due mainly to changes in level caused by movements of the earth's crust. Two factors are involved: subsidence and elevation; and the wear and tear by water, frost, and wind. An example of erosion is afforded by the coast of Suffolk between Southwold and Aldeburgh where within the memory of many of us the east end of the old Dunwich church stood several feet from the edge of the cliff. The church has been entirely destroyed by the sea with the exception of a piece of the tower that was removed and re-erected a short distance inland. The ruined church of Reculver on the edge of a cliff on the Isle of Thanet was nearly a mile from the sea in the reign of Henry VIII. It is said that between Flamborough Head and Spurn Point at the mouth of the Humber over a hundred square miles of land have been lost since the Roman invasion. Though it is not easy to distinguish between changes caused by the destructive action of the sea and those attributable to upward and downward oscillation of the land, it is often possible to obtain definite evidence of the latter. One example is worth giving: a Danish geologist took a series of photographs at intervals of a few years of the face of a cliff on the west coast of Greenland, selecting a spot which could be readily recognized by certain peculiarities; he found on comparing the photographs that the zone of brown seaweed adhering to the cliff and exposed at low tide had crept higher up the face. The land was slowly sinking or, it may be said, the sea was gradually rising. It is perhaps worth suggesting to younger readers willing to take a long view that they might start a series of similar records by photographing marked faces of an English cliff on a rocky coast.

Water freezes in cracks and fissures of rocks and expands approximately 10 per cent of its volume, chipping off blocks of various sizes. It is true we might watch the face of a crag for a long time without noticing any perceptible change, but the cumulative effect is apparent in the litter of loose pieces of rubble at its base. There is constant interchange and circulation of material between the atmosphere, the earth's rocky crust, and the oceans. Waste from the land is transferred to the sea both in a solid form, as sediment carried long distances by rivers and streams, and in solution. Continents and islands through the action of the weather—rain, frost, wind, and changing temperature causing expansion and contraction—are subjected to continuous disintegration: the products of destruction are the raw material of which new rocks are made. Geologists speak of weathering, or the effect of meteoric agents on the earth's surface, and of denudation, or laying bare (*denudare*= to lay bare) by removal of the loose debris, and erosion (*rodere*= to gnaw), the process of eating away. Rocks are broken into pieces; their surfaces are smoothed and pitted by the solvent action of water charged with acids derived from the air and from the ground, or reduced to loose, incoherent sand and particles of clay; the broken pieces and the finer material along with substances in solution are removed by wind or water, and the surface laid bare. Rivers erode channels and serve both as carriers of waste as well as cutting-tools. One can often see the effects of weathering on the pillars and walls of buildings and on tombstones made of certain kinds of rocks on which inscriptions are hardly legible. The effect of chemical weathering is particularly noticeable in districts where the ground is made of limestone, in West Yorkshire, many other parts of England, and in Ireland: fissures which cut the rock into broad strips and blocks are gradually widened by the solvent action of water and rendered habitable by ferns and other plants. Caves and underground watercourses are striking examples of chemical weathering. The soluble carbonate of lime removed from limestone by water containing carbon dioxide gas is often redeposited as stalactites and stalagmites as fur is deposited in pipes and kettles from hard water. Much of the dissolved material finds its way to the sea where it furnishes marine animals with the

PLATE I

Soil erosion and development of gullies on a deforested hillside

stuff of which their shells are made. Spray from a waterfall in a limestone district deposits lime on evaporation, and as the solvent power is reduced by decrease in pressure when water comes to the surface from depths where the pressure was greater, films of carbonate of lime cover the adjacent ground and in course of years may build up masses of a porous rock known as travertine, the material of which many of the old buildings in Rome are constructed. Travertine is widely used in modern buildings and can be recognized by the rather loose and porous texture made conspicuous on the polished surface.

Chalk is comparable to limestone in origin; it is made of fine particles of carbonate of lime together with innumerable fragments of often well-preserved specimens of shells of marine creatures; it was formed on the floor of a clear sea of moderate depth. A characteristic feature of chalk is the occurrence of irregular nodules of flint in rows roughly parallel to the planes of bedding. Flint is siliceous and was no doubt derived in part from the siliceous skeletons of sponges which passed into solution in the sea water and subsequently accumulated in patches of solid matter in the chalky ooze. The upper edge of a chalk cliff or railway embankment is often very uneven, and between the surface soil and the white rock is a layer a few feet in depth of brown rocky material made of flints and other insoluble residue left after removal of the soluble part of the chalk.

Plants play an active part as weathering agents. Soil consists of substances derived in part from the decay of animal and plant remains and in part from the disintegration of the underlying rock. By digging a hole in the ground we can trace a gradual passage from the surface-soil to the more gravelly subsoil below and deeper to the broken layer of solid rock. Roots of trees as they grow in girth exert a considerable levering force in cracks into which they penetrate, and the more slender roots aided by acids excreted from their cells co-operate with earthworms and countless hordes of microscopic organisms in reducing by mechanical and chemical means the coarser subsoil to the humus on the surface. The orange-coloured, black, and vermilion lichens, reflecting 'the sunsets of a thousand years', are out-posts of the plant world on inhospitable mountain tops which prepare the ground for mosses and flowering plants. Invisible bacteria with lichens and mosses all take a share in the dis-integration of rocks. But plants are by no means merely agents of destruction; they are among the most useful agents of rock-building. Generations of bog-moss and other plants which grow gregariously in wet and swampy ground and on moorland weave a thick covering of peat. Lakes and ponds are overgrown by the gradual advance of the marginal herbage: showers of vegetable debris from the plants floating on the water contribute to the black mud that lies below. To the accumulation of dead leaves, stems, and other scraps we owe both peat and many kinds of coal. An area long covered by forest may slowly subside and be overwhelmed by water which eventually spreads sheets of sand and mud over the buried vegetation: the dead forest refuse is the material of which coal was made, and the overlying sediment forms the hard rock that lies above a seam of coal. There are many ways in which plants contribute to the formation of rocks: fresh-water ponds and ditches are inhabited by a great variety of microscopic organisms among which Diatoms are abundantly represented; they are simple plants enclosed in a delicate case of almost indestructible silica. While most of the delicate plants are converted by decay into unrecog-nizable fragments the resistant protective coverings of the Diatoms remain and build up a white rock known as diatoma-ceous earth. On a small scale one can sometimes see such white layers in ground lying over the site of an old lake or pond. In Bohemia, California, Virginia, New Zealand, and other parts of the world there are thick beds of rock made almost entirely of the remains of fresh-water or marine Diatoms. In the Southern and Antarctic Oceans the sea-floor is covered for thousands of square miles with a light-coloured ooze consisting of the cases of millions of Diatoms, and we may safely assume that some of the hard diatomaceous rocks in the earth's crust are upraised sea-floors.

Another example of plants as rock-builders is afforded by certain seaweeds which cover their soft bodies with carbonate of lime and resemble corals, for which they were formerly mistaken: a few of these calcareous seaweeds are common on our beaches. Calcareous seaweeds are now abundant both in the Arctic regions and in warmer seas; in many coral reefs they outnumber the true corals. It has been found from an examination of many limestones of different ages that in the seas of earlier geological periods lime-secreting seaweeds and other organisms with calcareous skeletons lived under conditions similar to those that are favourable to the existence of their living descendants. Reefs now being formed in the Mediterranean Sea by the accumulation of the hard calcareous framework of seaweeds and animals provide clues to the interpretation of the conditions of formation of rocks which are a conspicuous feature in the Austrian Alps and other mountain ranges.

It is impossible to draw a clear distinction between the two aspects of what is in reality a continuous series of events, rock-construction and rock-destruction. The products of weathering and erosion are usually removed by water from their place of origin, but in certain circumstances they build up material *in situ*. An impressive example of the accumulation of material produced by weathering is furnished by the iron clays known as laterite, found in many tropical countries: the clays lie on the parent rock from which they are derived by chemical disintegration.

We will now turn our attention to the production in bulk of

loose material derived from coherent rocks. If we follow the course of a river from its source on the side of a hill through the lower reaches to the mouth we readily grasp the methods of erosion and denudation. The head of a mountain stream is strewn with blocks of rock detached by frost or by the battering of cliff faces by a torrent in flood hurling stones hither and thither: as the velocity and volume of the water increase larger and larger stones are swept along, pounding against one another and rolling over the rocky bed. Angular blocks are rounded into pebbles, broken into smaller pieces, into grains of sand and ultimately to particles of mud. The smaller and lighter the pieces the farther they travel suspended in the water. Here and there in a still pool rolled pebbles and sand come to rest, forming banks and layers, but most of the finer debris is carried to the main dumping ground where the flow is checked by contact with the water of a lake or sea. The sand no longer impelled forward sinks to the bottom, and the mud, lighter than the sand and more buoyant, floats for a longer period, the smaller particles coalescing into groups which slowly fall.

There are many stages in the manufacture of new rocks from old: the breaking and crumbling of rocks, the conveyance of detritus to lower levels, and in the course of the journey the gradual reduction of larger to smaller pieces by impact and grinding, the spreading of pebbles, sand, and mud over the reception area where the material is sorted by weight as it comes to rest. The next stage may be after an interval of many thousands or millions of years when through movements of the earth's crust the loaded floor of a sea or estuary is raised above the water and eventually, it may be, folded into arches and troughs or into fantastically crumpled masses that form part of a mountain chain. The soft material may be cemented into coherent sheets by the deposition of some binding substance introduced by percolating water before it is uplifted into a plateau or, if the uplift is less vertical and more lateral, into ranges and alps. Thus from river-deltas and sea-floors which were collecting grounds of river-borne sediment new lands rise from the water and the cycle begins afresh. It should be noted that when a river flows into the sea the sediment comes under the influence of currents which play an important part in its distribution. On many parts of the English coast we see shingle beaches between the edge of the land and the sand on the lower slopes of the shelving platform of the beach. Beyond the sand, samples dredged from the sea-floor show a gradual passage of sand to mud: these three kinds of deposits may have been derived almost entirely from the cliff overlooking the sea or they may consist in part of material that has been transported by rivers and currents far from its source. Cliffs are exposed to attack from the atmosphere and from the tides: the face of a cliff sloping backwards from the beach tells us that wastage of the upper part by rain and frost is more rapid than the wearing away of the base by stones on the beach used as battering rams by the waves. Other cliffs leaning outwards indicate a greater loss from marine than from subaerial erosion. There are other factors to be considered: rocks are not uniform in structure and equally resistant throughout; there are harder and softer layers; moreover the whole mass is more or less clearly intersected by divisional lines that may be horizontal, vertical or inclined at different angles; some of these are known as bedding-planes, that is, lines of partial separation of the rock into layers corresponding to the original beds or sheets of sediment; other divisional planes are known as joint-planes and were produced as cracks in parallel series under the influence of pressure to which the rocks had been exposed. These bedding- and joint-planes are contributory factors in the disintegration and wearing away of rock-masses.

A shingle beach well repays examination: the pebbles rounded and smoothed by the repeated backward and forward rolling together and attrition in the swish of the tide are not all of the same kind of rock; some come from the cliffs close at hand, others from more distant sources. It is often possible to trace them to their source either by noticing the nature of the rock or fossils they contain. If there is a sandy beach one can learn many things from it: samples examined under the microscope show that the grains are more or less angular; there is often an admixture of shell fragments and minerals other than silica of which most sand-grains are made. On the surface of a sandy beach or embedded in it are seaweeds, the dark egg-cases of fishes, shells, and feathery branched things which are often incorrectly called seaweeds but are the framework of colonies of very small animals known as Polyzoa which live in tiny cups grouped together in a flat, forked, or feathery framework. By taking note of the structure and shape of the sand-grains and of the animal and plant remains we can glean facts helping us to throw light on the method and conditions of formation of some of the rocks of past ages that were once sheets of soft sand. Pieces of wood and twigs of trees are not uncommon on a sandy beach, some carried by currents from places where they first reached the shore as flotsam from an inland source: a deposit found in a shallow sea may contain drifted samples of a land vegetation. There are many other things to be seen: the pulsating movement of the water transmitted to the mobile sand is recorded in curving ridges and grooves or ripple-marks, and these may be permanently preserved if they are gently covered by the next tide with a protecting layer of sand. Ripple-marks are often seen on the flagstones of pavements. Animals slithering or crawling over the wet sand leave various kinds of tracks and these too are preserved in hard rocks. Such markings on rocks have often puzzled geologists, as the animals which made them may be types unknown at the present day. The footprints of

the man Friday were a thrilling revelation to Robinson Crusoe:
we experience no such thrill from impressions made by boots
or bare feet, but we are more interested in the identification of
patterns made on the sand by dogs, rabbits, birds and other
wanderers. Large unfamiliar footprints made by extinct animals
are occasionally found on rocks that are now part of the land,
and from them it has been possible to learn something of the
creatures which made them. A shower of rain leaves scattered
crater-like pits, and if, as the shower fell, the water was driven
slantwise by the wind the unequal depth of slope of the walls
of the pit tells us the direction of the oblique impact made by
the drops. Rain-pits, a network of cracks made by the contraction of the wet sand under the warmth of the sun, and
ripple-marks on rocks formed millions of years ago serve as
guides to the interpretation of the past in terms of the present.

A sandy beach serves as an illustration of the behaviour and
eroding action of rivers: rills of water flowing over a gentle
slope form miniature channels, and tributary streamlets con-
verge towards the main channel where we can see grains of
sand and small stones rolled and carried by the water and,
finally, spread fanwise as a small delta. The branched channels
bear a close resemblance to the pattern produced when a small
stem with lateral branchlets is pressed against a yielding surface,
a deceptive resemblance that has misled geologists in their
observations on rocks of past periods. There are many pitfalls
in the path of those who try to read the story of the earth:
a piece of an old Wedgwood teapot was once described as part
of the stem of a plant that grew in the forests of the Coal Age;
a fused and rounded piece of rock from the crater of a volcano
was mistaken for the inflorescence of a tropical aroid. One
often sees in private collections of miscellaneous objects delicate
branched markings on the surface of a flat stone believed to be
fossil mosses; they are patterns made by a mineral substance

which simulates plant forms. Mistakes are unavoidable even by experienced geologists who are sometimes taken in by puzzles and traps set by Nature in a mischievous mood.

Sand is now being accumulated not only on gently sloping shores but at many places also as dunes or small hills piled up by the action of wind. A dune begins as a small heap of sand-grains driven along the ground against some obstacle and is gradually increased in height as fresh supplies convert the mound into a hillock with its surface fashioned like driven snow into beautiful, curved contours. Sand-dunes are seen in fullest development in desert countries and on many shores where disintegrated fragments of rock become the sport of the wind. The grains of which dunes are made are usually more completely rounded than those on a sea-beach owing to the constant move-ment of the sand on the wind-swept mounds. Shells of land-snails and remains of other animals occur in dunes along with plants that are able to grow under the relatively dry conditions characteristic of highly porous ground and of localities where water is scarce. By the nature of the material as well as by the fossils, or by the general barrenness of the rock and absence of fossils, it is often possible to recognize from rocks in the earth's crust evidence of an environment similar to that in which dunes are now formed.

Mud flats of estuaries and muddy sediment on the sea-floor farther from land than the sandy beach consist of layers of finer material that has been transported farther from land by reason of its lightness. A shelving floor beyond tide-level is inclined at a regular and low angle to the horizontal in zones parallel to the coast—pebbles of the shingle beach, sand in layers tapering toward the deeper water and farther away a broad belt of muddy sediment. We know that the land is slowly but surely being worn down by the instruments of erosion and denudation and, in course of time, the surface inequalities will tend to be reduced almost to a level plain or rather to a plain sloping gently towards the sea. In the process of planing down mountains and hills the sea plays its part. The shelving plain may be covered to a depth of many feet by sand and mud or, as can be seen on many parts of the coast, it may be smoothed into a rocky platform with pools where sea anemones, other marine animals and seaweeds find a congenial home. In some regions planed-down surfaces of the land are represented not merely by the present platform sloping away from the coast-line, but are recognizable far inland and at a higher level. If one stands on an eminence commanding a view of ranges of hills, one behind the other, it is sometimes possible to detect an approximate approach to a general uniformity in the height of the more prominent features in a Lakeland or Highland land-scape. A line drawn over a wide area passing through the highest points follows a gently sloping direction, and this suggests that we are looking at an old plain of marine denuda-tion; or, in other words, the surface features which are the expression of the cumulative and long-continued work of sub-aerial eroding agents are subordinate to a much wider feature represented by the imaginary line drawn over the crests of the hill ranges. The land was once overwhelmed by the sea which gave the finishing touches by planing it down to what is called a plain of marine denudation. Such evidence of the part played by the sea in wearing down the land over which it had trans-gressed enables us to picture a far-off scene before the hills and valleys had been fashioned by rain and frost, a scene where the land lay under water as a shelving platform that was later lifted up to a higher level as the sea retreated. The emergent land at once came under the influence of eroding agents that carved it into hill and dale, and yet the signs of the old plain of marine denudation were not wholly obliterated.

Beyond the outer edge of the shelving platform off the coast, with its continuation as the continental shelf under fifty to a hundred fathoms of water, and beyond the reach of all water-borne sediment from the land the sea-floor is covered with other kinds of material derived from myriads of creatures whose home is in the sea. In many parts of the world where the temperature of the sea is not lower than about 60° F. animals closely related to sea-anemones secrete from the water carbonate of lime and use it to form a hard framework or internal skeleton in and around their soft bodies, part of the material of which coral reefs are made. The living coral polyp builds on the dead foundations laid by its predecessors and masses of calcareous rock are gradually formed as fringing or barrier reefs or as coral atolls. Coral reefs are not built exclusively of coral but in part of the remains of lime-secreting seaweeds. Over many thousand square miles of ocean-floor, at an average depth of about 2000 fathoms in warm temperate and tropical regions, there is slowly accumulating a chalky kind of white mud known as Globigerine ooze because it contains millions of shells and fragments of shells of minute marine animals belonging to the class Foraminifera, so called because the shells encasing these very lowly organisms are perforated by small pores, of which *Globigerina* is one of the commonest representatives. The shells of Foraminifera and other animals living on or near the surface of the water fall to the bottom and contribute to the formation of a calcareous deposit, along with much larger shells of molluscs, oysters and many other creatures and in some seas the framework of corals. Our chalk downs and limestone hills are made in great part of the miscellaneous shells and other hard parts of innumerable marine creatures which lived in ancient seas in water uncontaminated by sediment carried by rivers and currents from the land. Deposits now being formed in modern seas, though not identical in the remains of plants and animals with those uplifted from seas of former ages, agree closely enough to help us to reconstruct the conditions and manner of their formation. It would not be correct to describe chalk or limestone as a rock entirely composed of shells and corals: in addition to recognizable traces of animal and plant remains, there is other material made of fine particles of carbonate of lime and allied substances which had a different origin. It has been found, in the Bahamas Sea and elsewhere, that large quantities of a fine-grained calcareous mud are being deposited, and that this mud is due to bacteria and certain other microscopic plants, whose activities cause the precipitation of limy substances. In all probability we have here another instance of the applicability of present methods of rock-formation to the better understanding of what occurred in the past.

In this brief and incomplete sketch my aim has been to stimulate observation and to awaken a spirit of enquiry rather than to describe in detail what can be gathered from geological text-books. The next chapter is devoted to rocks, some of which are strictly comparable with rocks now in the making, while others have no modern counterpart accessible to us.

CHAPTER IV

Rocks

In ordinary parlance the term 'rock' is applied to hard stony material such as sandstone, grit, limestone, granite, basalt and many other constituents of the solid part of the earth's crust. Writers of geological text-books use the term in a more comprehensive sense and define a rock as an aggregate of minerals whether coherent and relatively hard or incoherent and of loose texture such as gravel, sand, clay, and mud. The more compact rocks differ in hardness: chalk and limestone can be scratched with a knife and effervesce when a drop of hydrochloric acid is put on to the surface; on others, such as sandstones, the blade of a knife makes little or no impression, neither does hydrochloric acid. The earth's crust is made of two classes of rock differing in composition and in origin, sedimentary and igneous. Sedimentary rocks are widely distributed and cover perhaps as much as 75 per cent of the whole surface; they are mainly old sediments that were once masses of gravel, sand, and mud transported by water and deposited in layers, sheets, and banks in estuaries, lakes, and seas. A few sedimentary rocks were formed on land: some are old surface-soils, e.g. the bed of fire-clay that often lies below a seam of coal and is the altered surface on which forests grew; some contain a high percentage of carbon, e.g. coal and allied material, and consist of a compacted and more or less chemically altered accumulation of vegetable debris; others were formed without the intervention of rivers, as masses of broken rocks that had been talus dumps on the steep flanks of mountains, such as we see now on the shore of Wastwater and in many other places at the foot of a cliff, or as piles of loose sand blown by the wind into dunes. In the second class are included many varieties of rock all of which were produced not by destructive eroding agents but by the action of very high temperatures. Igneous rocks were originally liquid or molten and solidified as the temperature fell below melting-point: their texture—whether glassy, or finely or coarsely crystalline—varies according to the circumstances in which they became solid. Crystalline igneous rocks, and they are by far the commonest, consist of an intricate mosaic of interlocking crystals, a structure which makes them tough and hard; they differ widely from sedimentary rocks that are composed of mineral particles either large or small which were compacted and pressed together into more or less coherent layers from masses of loose and independent grains.

SEDIMENTARY ROCKS

SANDSTONE AND GRIT

It has been estimated that if the sedimentary rocks of all geological periods could be arranged as a single pile it would be 50 or 60 miles high.

A sandstone as the name implies consists mainly of a comparatively coherent mass of sand (chiefly silica), the grains being cemented together by some binding material introduced by water percolating through the porous sediment or derived from the partial solution of the grains themselves. Some sandy rocks are not hard and compact but consist of loose sand which has retained more or less unaltered the texture of the sediment as it was first deposited from water. Examples of old beds of sand that have not been converted into hard rock are particularly common in English counties where the surface is made of sedimentary deposits belonging to the Tertiary era. In many parts of East Anglia, especially on the coast near Southwold and elsewhere, in the sandy heaths and woods of Surrey, and in the cliffs in the Bournemouth district, loose masses of sand are a conspicuous feature. Most sandstones are hard though they differ in the degree of union between the individual grains; a coarse-grained sandstone is usually called grit. In red sandstone and grit the colour is given by a thin film of an oxide of iron (comparable with rust) that coats each grain. Sandstones and grits, as also other sedimentary rocks, are usually arranged in layers varying in thickness from a few inches to several feet: this layered structure—the regular bedding of the rock—is an expression of the manner of origin; the material settled on the floor of a shallow sea or estuary as sediment and after a pause fresh supplies were thrown down. The interval between successive loads of sand is represented by a line of weakness, the junction between a partially dried and compact surface and fresh layers of looser sediment; along the dividing plane the rock can be easily split into flags suitable for paving-stones, or into large blocks. If we look at an exposure of rock in a quarry the bedding-planes are clearly seen, but one sees also other lines of division at right angles to the bedding: these are joints (Chapter III, p. 35 and Plate II, upper figure) that constitute additional lines of weakness and greatly facilitate quarrying. There may be two sets of joint-planes at right angles to one another which together with the bedding-planes divide up the rock completely. By drilling a series of holes through the rock and splitting it by inserting an explosive charge into each hole or by other means the tendency to split along bedding- and joint-planes makes it easy to detach fairly regular, rectangular blocks. Joints are particularly well shown in limestone that forms the surface of a plateau; they form an irregular network of gaping fissures in which ferns and other plants are abundant. In origin, joints differ fundamentally from planes of bedding; they were produced by pressure caused by movements in the solid crust which compressed the rocks

PLATE II

(*a*) Bedding and joints
Limestone and shale of the Lower Lias, Penarth

(*b*) Perched block, Norber (see p. 105)

and subjected them to severe strain. The mass under pressure cracked along regular planes in response to the disturbing forces. If a piece of glass is fixed at one end and the other end bent upwards and downwards it cracks along roughly parallel lines which are often in two series at right angles to one another. It is well known that the earth's crust has repeatedly been exposed to lateral pressure causing contraction and strain, and one effect of this is seen in joints.

Before leaving sandstone and grit it is worth noticing a fairly common feature which throws light on the conditions under which the rocks were deposited. If we look at some sandstone exposed in railway cuttings or large weathered blocks that are scattered over moorlands we can often see series of roughly parallel lines rendered conspicuous by long exposure which occur in series oblique to the main lines of the regular bedding-planes: there may be several series inclined to one another at varying angles. These lines represent what is known as current-bedding or false-bedding and are evidence of deposition of the sediment by shifting currents or eddies in shallow water near a coast, or by wind along the sloping surfaces of dunes.

OLD SHINGLE BEACHES OR CONGLOMERATES

A conglomerate is a rock made of rounded pebbles usually firmly embedded in some cementing material; it is exactly like a modern shingle ·beach except that the pebbles have been cemented together. Rocks of this kind afford valuable clues to the positions of old shore-lines, and the pebbles of many different sorts help us to form an idea of the composition of the rocks which were the source of the conglomerates, grits, and sandstones. Beds of conglomerates and other shallow-water sedimentary rocks may be several hundred feet in thickness and this is an indication of formation in a basin of deposition which was slowly sinking.

BRECCIAS

The term 'breccia' is given to rocks consisting of a heterogeneous mass of angular and not rounded pieces: they differ
from conglomerates in the absence of well-rounded and water-
worn pebbles and are comparable to accumulations of broken
rock fragments on talus slopes or to collections of debris swept
from gullies and river-channels by torrents caused by rare rain-
storms in a dry region and not carried far enough to be ground
into rounded pebbles. Some breccias are volcanic in origin and
are made of fragments of rock that were exposed to explosive
forces released during eruptions.

CLAY, SHALES, AND SLATES

Shale may be described as a hardened and consolidated
thinly bedded clay or mud made of fine particles derived from
the mechanical and chemical disintegrations of mineral substances in rocks that had been exposed to the action of eroding
agents. Shales are very often associated with grits and sandstones; they indicate deposition of finely divided sediment
rather farther from land than the places where the heavier and
coarser sand was deposited. A shale is easily broken and tends
to split along the planes of bedding: it has an earthy, clay-like
smell. Fossils, both animals and plants, are common in shaly
beds and often well preserved; in some instances shells retain
their pearly lustre. As we shall see later, deposits of clay are by
no means always in the form of relatively hard, layered shale;
there are thick masses of clay belonging to many geological
periods which have undergone little change since they were
first deposited on old sea-floors, e.g. Kimmeridge Clay, so named
from Kimmeridge in Dorset, Oxford Clay, Gault and London
Clay.

Slates are harder than shale, and the more typical examples
that are used for roofing can be split almost indefinitely along

parallel planes, frequently at a high angle to the bedding: this property, known as slaty cleavage, has been impressed upon the rock by pressure generated in regions of the earth's crust subjected to forces of compression associated with crustal disturbances. Under great pressure rocks become semi-plastic, and particles such as small flakes of the mineral mica are able to move and alter their position; they arrange themselves at right angles to the direction of the pressure and that imparts to the rock a texture which allows of splitting along planes of least resistance due to the parallel and regular orientation of the minute particles. Bands of different colours are noticeable features on the smooth cleavage-surfaces of many slates, purple or dark blue in colour: the bands mark the layers of deposition or sedimentation which may be at various angles to the cleavage-planes. Slates have been formed in different ways: some are altered sediments transported as clay or mud by rivers that held in suspension material made from the scouring and disintegration of rocks; some consist of finely divided volcanic ash, that is, the material which is frequently thrown in dense clouds from craters within or below which violent explosions have shattered the adjacent rocks into fine dust. It was volcanic dust that buried Pompeii in A.D. 79. This product of explosive eruptions is not confined to the land; it may be spread by water over an ocean-floor from centres of activity under the sea. It is important to remember that the term 'slate' is also applied to rocks that are neither shale nor slate in the strict sense; slabs of a yellow rock used for roofing houses in the Cotswold country and some other districts are comparatively thin beds of a sedimentary rock that is mainly composed of calcareous and not clayey material. Stonesfield slate (from Stonesfield in Oxfordshire) and Collyweston slate (from the Stamford district) are examples of the application of the term to a rock because of its use for roofing purposes and not for a geological reason.

CHALK, LIMESTONE, AND OTHER CALCAREOUS ROCKS

Reference has already been made to chalk and limestone as rocks characteristic of clear water as distinct from sand and clay, the sediments deposited in a relatively shallow sea or lake and transported by rivers and currents from the neighbouring land: the calcareous sediments accumulate where they are produced, they occur in the place of origin and are made in part of shells and skeletons of animals, occasionally also of the calcareous coverings of certain seaweeds which lived in the sea. If beds of sandstone, shale, and limestone are found in the face of a cliff or at different levels on a hill-side lying one above the other, they furnish evidence of a sea-floor increasing in depth from a shallow sandy beach to rather deeper water with a muddy sediment and to a still deeper sea out of reach of river-borne sand and mud.

Limestones may be also fresh-water in origin; but most of them contain marine fossils which are proof of formation in salt water. A limestone may be produced by deposition of calcareous material from fresh water, holding in solution carbonate of lime obtained from limestone rocks over which a stream or river has passed. The face of a cliff overlooking the river Nidd at Knaresborough is coated with a white deposit of lime thrown down on the evaporation of the spring water. The so-called petrifying springs at Knaresborough and Matlock are examples of the formation of calcareous rock as the result of evaporation of water rich in dissolved carbonate of lime. This is the porous rock made up of thin calcareous layers known as travertine; it was used by the Romans in the construction of the Colosseum and other buildings; and the same rock is widely employed with good decorative effect in modern English buildings. Similar calcareous deposits are familiar as stalactites and stalagmites in many caves in limestone districts.

OOLITES

The name 'oolite' is given to some limestones made wholly or in part of small, spherical grains resembling fishes' roe (Greek, *oon* = egg); the grains can be seen with the naked eye, though much more clearly with a pocket lens; they are particularly striking in the yellow and occasionally pink Ketton stone named after Ketton in Rutlandshire; they are also a characteristic feature of rocks from other localities, e.g. the stones used in many of the Oxford colleges. Oolitic grains are made of concentric layers of carbonate of lime and there is often a nucleus of some sort in the centre, a tiny piece of shell or a sand-grain, around which lime was deposited in even layers as the growing oolitic grain was carried hither and thither by currents in the sea.

MARBLES

'Marble' is a name commonly given to a rock which takes a polish and is suitable for ornamental building or decoration; the term as used in commerce has no precise geological significance. Many commercial marbles are not sedimentary but igneous in origin. Limestone is a good illustration of a marble that is sedimentary: several limestone rocks of widely separated geological ages are used as marble mantelpieces, or small columns in churches. It is easy to recognize the age and the provenance of some of the limestone marbles by the fossils rendered conspicuous by polishing: mantelpieces are often made of dark grey limestone belonging to the Carboniferous period and quarried in Derbyshire, Yorkshire and other districts containing fossil corals, sea-lilies, and shells; limestones of Devonian age from the neighbourhood of Torquay and other places when polished show beautifully preserved corals characteristic of the period; another well-known limestone, Purbeck marble, was first used in the thirteenth century for shafts, or comparatively

SG 4

slender columns, in cathedrals; it is light green and contains many shells which denote a fresh-water origin. In geological parlance the term 'marble' is confined to limestone which has become crystalline owing to the action of heat, or heat and pressure combined.

SEDIMENTARY ROCKS OF VEGETABLE ORIGIN

Of these the most important is coal, and a description of it will be found in Chapter XIII. There is a gradual transition between rocks such as some shales which are almost black in colour, because of the abundance of carbon derived from plant remains mixed with the muddy sediment, and beds of coal in which plant debris almost completely takes the place of clay or sand.

It is hardly necessary to point out that it is impossible in many instances to draw a definite distinction between the several kinds of sedimentary rocks: one kind shades into another. A sandstone in one locality is made entirely of grains of silica; in another it is partly sandy, partly argillaceous, that is, there is an admixture of clay; a sandstone may also contain bands and streaks of a coaly substance furnished by plant remains. Such terms as sandy, or arenaceous; shale, argillaceous, or clayey; limestone; carbonaceous sandstone, etc., are applied to intermediate varieties which must necessarily occur in Nature.

IGNEOUS ROCKS

Igneous rocks must have preceded those made of old sediments —gravel, sand, and clay—which presuppose the existence of a source from which such material is produced by erosion and denudation. Molten and semi-molten lava flowing sluggishly from the crater of a volcano, comparable with the slag of a blast furnace solidified from a fused mass, is a modern example of the kind of rock that we associate with early stages in the development of the earth's surface. Lavas are now being added to the land, and from still active volcanic islands such as Stromboli, the 'lighthouse of the Mediterranean', as well as from centres of eruption hidden under the sea, lavas and other products of eruptive forces contribute to deposits on the ocean-bed. There are other igneous rocks of which there are no modern counterparts that we can see in process of formation: coarsely crystalline rocks, such as granite, passed from a molten to a

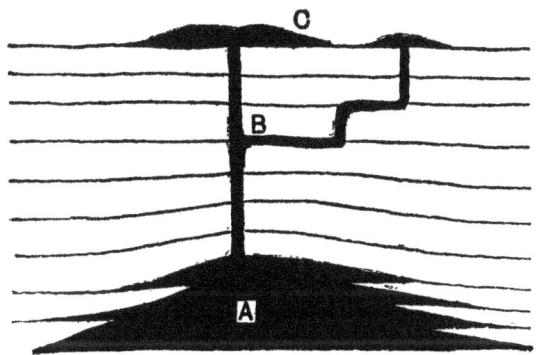

Fig. 2. Igneous rocks. *A*, Plutonic rocks; *B*, dyke rocks; *C*, volcanic rocks or lava.

solid state far below the surface, and it was only after long exposure to destructive agencies and the removal of the original cover that they have been exposed to view. We cannot tell whether such rocks are now being forced upwards from subterranean reservoirs far beneath our feet.

Igneous rocks are conveniently divided into two not very sharply defined classes, Volcanic and Intrusive; the more deeply seated intrusions are known as Plutonic (from Pluto, the god of the underworld) (Fig. 2). Both kinds were originally molten and the difference between them is the result of the conditions under which they became solid and crystalline. Among rocks associated with volcanic activity are lavas of various kinds; some did not cool slowly enough to permit the development of crystals in the molten liquor but solidified rapidly as volcanic glass. A good example of volcanic glass is the great sheet of pitchstone, 400 ft. thick, which dominates as a steep-sided, sloping cliff the island of Eigg in the Inner Hebrides. Microscopical examination of transparent slices of a glassy rock reveals the presence of minute beads and needles that represent initial stages in crystallization within a homogeneous matrix. Lavas as a rule are not glassy in texture but consist of crystals differing in chemical composition that are too small to be seen with the naked eye: a lava-flow has a slaggy and irregular surface, and the uppermost part, which has cooled rapidly on exposure to the air and has a frothy porous structure caused by escaping steam and gases, solidifies as pumice. Lavas are found among rocks of many geological ages, and one of the commonest is known as basalt, a dark grey or black, finely grained rock which becomes brown on exposure to the weather. Basalts mainly occur as old lava-flows poured over a land-surface or over the ocean-floor; some are intrusive rocks that did not reach the actual surface but were forced upwards as sheets and sills of molten material along bedding-planes, lines of weakness between one layer of rock and another; the intruded sheets were quickly cooled by contact with the surfaces of the invaded beds. Basaltic rocks, especially intruded sheets, vary in the size of the crystals and within certain limits in chemical composition; distinguishing names given to different types of basaltic rocks need not be mentioned here. Basalts and allied rocks cover many thousand square miles of the earth's surface in several regions where they have been poured out as lava streams or as fiery floods from fissures. An impressive example of this is described in Chapter IX. The basalts of the Giant's Causeway in northern Ireland and at Staffa and other islands off the west coast of Scotland furnish illustrations of a common structural feature, the division of the rock into hexagonal columns due to contraction on cooling and analogous to the columnar form assumed by starch on solidification.

Basalts also occur as dykes, which are sheets of rock forced in a molten state into vertical or oblique cracks in the earth's crust. Dykes being frequently harder and more resistant than the rocks into which they were intruded stand out as conspicuous walls several feet in thickness above the general level: on the other hand, as can be seen on the beach of the island of Arran, dykes occasionally occupy trenches in the penetrated rock. There are at least four hundred dykes in the Isle of Arran varying in breadth from a foot to a hundred feet: each of them marks the position of a rift in the rocks which they invaded and that means a stretching of the earth's crust. It has been calculated from the total thickness of the dykes that the crust was stretched more than five thousand feet.

One of the most remarkable examples of a basaltic rock that is not a dyke but a sill or intrusive sheet is the Whin Sill; it was forced from a subterranean source into an overlying series of sedimentary rocks belonging to the Carboniferous period and this probably occurred in the Permian period. The sill extends across England, sometimes hidden, in some localities forming an escarpment or other conspicuous feature, from Westmorland to the Farne Islands which, together with the cliffs at Dunstanburgh and Bamburgh, are built of the basalt; it is a hard bluish grey rock, 80 to 100 ft. thick, lying as a regular sheet intercalated among the beds of Carboniferous sediments and occasionally passing from one bedding-plane to another and altering the texture of the rocks between which the molten intrusion squeezed its way. Being harder than the adjacent rocks the Whin Sill has been worn away less rapidly by water and left as a scarp causing a sudden descent of streams that pass over it as waterfalls, e.g. High Force in Teesdale and several others. Near Greenhead in Northumberland, where the rocks that once lay above it have been removed by eroding agents, the Whin Sill stands as a bold escarpment rendered all the more impressive by the Roman wall which pursues its undeviating way over the crest.

Rocks included in the plutonic category differ from basalt and other finely grained products of igneous activity in their much more obvious crystalline structure; they were not poured over a land-surface nor did they solidify near the surface but under a heavy load of rocks above them: they cooled very slowly and under conditions favourable to the formation of large crystals visible to the unaided eye. The best and most familiar example is granite, a pink or grey rock made for the most part of three easily recognizable minerals, felspar, mica, and quartz (silica): felspar is by far the most conspicuous and occurs as grey or pink crystals, sometimes more than an inch in length. Different kinds of crystals, varying in colour and composition, are included in the felspar group. The mica crystals are much smaller and split readily into thin glistening flakes when gently pressed by the point of a knife. The interstices between the felspars and mica are filled with less regular patches or blebs of quartz too hard to be scratched with a knife. As the molten material slowly cooled the mica crystals separated out first, then the felspar, followed by quartz. Pink granite quarried at Shap in Westmorland is extensively used as polished slabs on the face of buildings and for columns and pedestals: the grey granites of Aberdeen have been used in the construction of many buildings in Aberdeen itself and elsewhere. One often sees on shop fronts and on the façades of other buildings polished slabs and blocks of a purplish blue plutonic rock of Norwegian origin known as syenite; it has large, attractively iridescent crystals.

Another plutonic rock, related to basalt and distinguished by its relatively large crystals, is gabbro, the rock mainly responsible for the formation of the Cuillin Hills in the island of Skye. Gabbro differs from granite in its dark, almost black colour, and in its mineral composition. Further reference is made in later chapters to the Cuillin Hills and to other masses of plutonic rock that play a part in determining scenic features. All plutonic rocks have not been produced on the same plan and in the same way; many of them may be regarded as masses of molten material impelled upwards from a subterranean source which, failing to reach the surface by penetration of the rocks blocking their path, gradually passed into a solid coarsely crystalline form below a thick covering of the crust. Their formation may be described in rather crude terms as follows: some powerful force within the deeper regions caused large quantities of molten rock to rise to a higher level; the material as it rose pushed upwards the rocky covering into a solid bubble or blister or more correctly into a dome, and finally crystallized; later on, weathering agencies gradually removed the greater part of the domed layers and at length laid bare the initial cause of the uplift, thus exposing to our view plutonic rocks that were once hidden.

METAMORPHIC ROCKS

Geologists make use of the diversity of structure of rocks as a means of reconstructing past events: sedimentary rocks agreeing very closely with sediments now being deposited tell a straightforward story; others, especially in the older parts of the earth's crust, are more difficult to interpret. Reference has already been made to rocks that cannot be closely matched with any that are now in the making. There is another category of special interest to geologists mainly concerned with rock-structure and of general interest to the amateur: when molten matter is thrust

upwards into rocks lying in its path the intense heat and force
of impact produce a more or less profound alteration in structure
and texture. The same kind of effect is caused by heat generated
when rocks are folded and crumpled. Rocks altered by heat are
metamorphosed, i.e. they are changed into something different
from their original state. Limestones are changed into crystalline
marble, sandstones into quartzites; shales into schists, and so on.
Rocks abutting on dykes and sedimentary beds in contact with
sheets of basalt or other igneous rock that had squeezed its way
along paths of least resistance are found to have undergone
varying degress of metamorphism. Enough has been said to
show how difficult it is to distinguish between structures that
are original and characters superinduced by metamorphism.
Metamorphic rocks include both sedimentary and igneous
classes; it is only in recent years that modern methods of technique have enabled geologists to appreciate the transforming
effects of agents of metamorphism.

A coarsely crystalline metamorphic rock, especially noteworthy as a very widely spread constituent of some of the oldest
parts of the earth's surface, is gneiss, very similar in composition
to granite but distinguished by a more or less regular banded
structure, grey, pink, and dark layers in which the crystals are
easily seen. Some gneisses are igneous in origin; others are
highly altered sedimentary rocks, and it is difficult to draw a
line between the two. Gneisses that are truly igneous and not
transformed sediment owe their banded structure to the arrangement of the component minerals in layers when the rock was
plastic enough to permit movement and fluxion in response to
pressure. The name 'schist' is given to a finer-grained crystalline
rock with a tendency to split into layers, frequently characterized
by glistening surfaces reflecting light from crystals of mica.
Many schists are altered sedimentary rocks; they occur in
districts where there is evidence of violent disturbance and

crustal folding. Examples of gneissic and schistose rocks are given in Chapter XVII.

The natural history of rocks, especially those that were once a liquid mixture of silicates far below depths accessible to investigation, is a branch of earth-history demanding a knowledge of chemistry and physics. Readers who wish to know more of this fascinating line of enquiry should consult books devoted to Petrology, that is, the study of rocks.

CHAPTER V

The Distribution of Rocks over the Surface of Britain

Travelling through the length and breadth of England by rail or road we see comparatively little of the rocky floor except in districts where hills and ravines are conspicuous features; hundreds of square miles are covered with grassland with here and there woodland, moorland, and fen. On closer observation some indication of the nature of the foundation-stones can be discovered; the colour of the soil may serve as evidence of the kind of rock that lies below the surface, scars on grassy slopes, scarps overlooking a river valley, layers of rock in a railway cutting, in quarries and other exposures, and the changing character of the scenery. From these and other sources of information it is possible to learn much about the solid substratum. On a geological map of the British Isles the whole area is divided into strips and patches, with distinctive colouring to indicate the different periods, which show the distribution of rocks over the surface of the ground. The map teaches us much more than can be seen either from the air or in journeys on land: the rocks are partially hidden not only by cultivation but by the frequent occurrence of what are technically called superficial deposits consisting of widely distributed accumulations of gravel and sand and ridges and mounds of stiff clay containing a miscellaneous assortment of boulders, deposits that are the legacy of the Great Ice Age (see Chapter VII). Two sets of maps are published by the Geological Survey, one of which shows the superficial deposits including the comparatively modern sheets of clay and other sediment left by rivers on land bordering their lower reaches, material marked on maps as alluvium; the other series of maps omits the superficial deposits and shows the rocky floor below.

Making an accurate map of the more solid substratum is by no means an easy matter: an exposure in one locality may be separated from the next by a mile or more of intervening country where the rocks are hidden. It is however legitimate to assume continuity between the two exposures even though the connecting rocks cannot be seen if those actually observed are clearly of the same age as shown by the fossils or the nature of the rocks. In the map reproduced (Fig. 3, p. 61) we see the rocks of different periods in surface-view, their distribution in relation to one another; but that is not all the information given by the map. It is not enough to know how the sedimentary and other rocks are arranged at the surface; we want to know their positions relative to the surface, whether horizontal or inclined at a greater or smaller angle. If we look on the map at the Pennine Chain lying in a north and south direction in Lancashire, Yorkshire, and Derbyshire we notice a repetition of the same set of rocks on both flanks. This repetition gives a clue to the inclination, or dip, of the rocks and confirmatory evidence is supplied by exposures on hill-sides, quarries, and river valleys. The central axis of the Pennines is built of layers of limestone, well displayed in the Derbyshire dales, which form a broad arch with younger sedimentary beds lying on the sloping sides on the east and west. The younger beds above the limestone are now confined to the two flanks, but they were once continued over the central arch and have been removed by long-continued erosion. See Fig. 8, p. 211.

In some regions the original position of rocks that were deposited as horizontal or slightly inclined beds of sediment has been almost completely retained: great blocks of the earth's crust have been lifted up without being folded and now lie in regular layers of Nature's masonry below tablelands and prairies. Much valuable information on the underground disposition of rocks is furnished by deep boring from which cores drilled by cutting tools can be examined. Bores sunk in the London area reveal the existence of a very old folded ridge that, in the remote past, was a range of mountains stretching across the southern part of England and beyond to the Continent. This hidden relic is all that remains of a corrugated rib of Palaeozoic rocks once a prominent feature in the landscape: after its elevation from an ancient sea it was exposed to erosion and eventually sank below the waves where the denuded roots of the hills served as a platform for the piling up in later ages of the chalk and other rocks that in their turn came to be parts of the land.

How, it may be asked, is it possible to refer rocks seen at the surface to their respective places in the geological table? The answer is, by observing the relative position as seen in exposures along cliffs by the sea, in river valleys and elsewhere; also by comparison of fossils collected from the sedimentary beds. As a general rule the orderly sequence of rocks as seen in the face of a cliff is a safe guide to age, the lower beds being the oldest. There are, however, notable exceptions: one of these, furnished by rocks in the Scottish Lowlands, is briefly described here because it serves as a particularly instructive illustration of geological mapping in a district where Nature seems to have set a trap for men who seek to pry into its secrets. Before giving a much simplified account of this example of complicated structure a short digression may be usefully made on the value of certain kinds of fossils as guides to the determination of geological age. Some animals and plants are known to have persisted with comparatively little change through several periods; others endured for a short time and vanished: the latter, short-lived types are often spoken of as index-fossils because their occurrence in a particular layer of a thick mass of

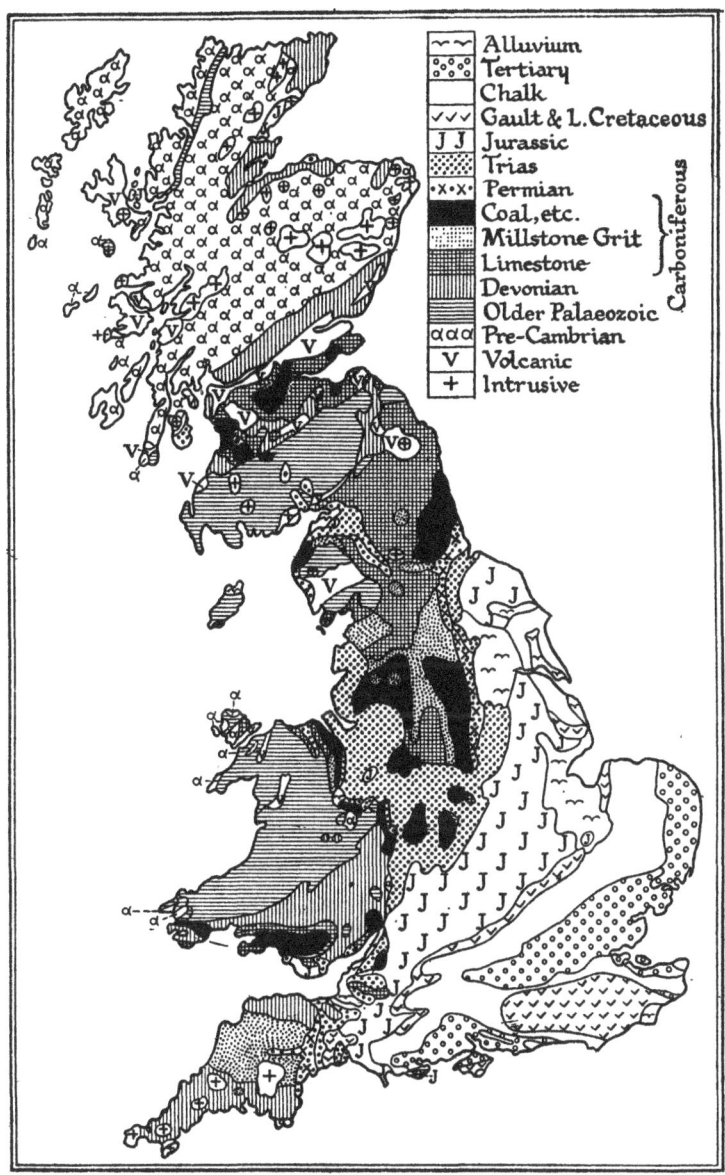

Fig. 3. Geological map of Britain.

rock serves to date the bed in which the fossils are found. Two examples may make clearer the employment of fossils as time markers. In England and northern France the Chalk reaches a thickness of several hundred feet, representing an enormous lapse of time. For purposes of correlating exposures of so thick a rock it is important to have some means of ascertaining whether we are looking at parts of the older, middle or newer stages: in other words, is it possible to distinguish horizons? Sea-urchins are abundant as fossils in the Chalk and often well enough preserved to enable an expert to distinguish with confidence one species from another. Several years ago an English doctor, A. W. Rowe, took up the study of fossil sea-urchins as a hobby and collected enormous numbers at many localities; he found that species of a genus called *Micraster* are confined to certain well-defined levels or horizons and therefore could be used as trustworthy index-fossils. When he found *Micraster* species A he was able to say that this layer could be described as the zone or level conveniently called *Micraster* A (the specific name is omitted). Thus by means of different species of this genus he was able to compare the age of the rock in widely separated districts. Another example is furnished by fossil species belonging to a wholly extinct group of marine animals, the Graptolites—so called from their resemblance to marks made by a lead pencil on slate (Greek, *grapho* = I write). Graptolites are characteristic fossils in some of the older Palaeozoic sediments; they are relatively simple in construction and may be compared with members of a living group known as Polyzoa, so named because they are colonial animals in the sense that numerous minute individuals live together in a supporting framework made of a chitinous or horny material. Certain kinds of Polyzoa common on sea-beaches are often mistaken for feathery sea-weeds, light brown in colour and a few inches long. Graptolites, like Polyzoa, were colonial animals. Species lived but a short time and the fossil remains have, therefore, a short vertical range in the rocks containing them. In the Scottish Lowlands in the region north of the Solway Firth extending inland from the coast near Ballantrae there are many highly inclined beds of shale, that is, old muds from the sea-floor: these beds as seen in exposures apparently form a perfectly regular series arranged in the order of relative age, the beds *a*, *b*, *c*, *d*, *e* seem to be in sequence, *a* being the youngest. When graptolites were collected it was discovered that certain species occurred in bed *a*, then disappeared and none were found in beds *b* and *c*, where their place was taken by other species; but in bed *d* the species of bed *b* reappeared, and in bed *e* those of bed *a*. This recurrence of graptolite faunas was worked out more than sixty years ago by Professor Lapworth of Birmingham, one of England's ablest and most original geologists.

He showed that the reappearance of the faunas in inverse order was a tectonic and not a depositional phenomenon, due to overfolding with part of the strata inverted. The zoning of the older Palaeozoic rocks by the same methods as have been used in newer rocks was thus made possible and used to unravel highly complicated geological structures.

In the chapters devoted to descriptions of geological periods many examples are given of folding of the earth's crust. The regular succession in parallel series of layers of rock inclined at varying angles is the result of upheaval of part of the earth's crust which disturbed the originally horizontal beds: when upheaval was less gentle the crust was folded into regular, symmetrical arches and troughs; when the pressure was greater the folds became steeper and the limbs of the arches and troughs asymmetrical. If a pile of coloured layers of cloth is pressed down by a weighted board and the ends forced towards one another the compression causes the layers to fold into arches and troughs. Another method which has been described illustrating the result of lateral compression is to spread dry stucco powder in thin layers between thicker layers of damp sand of different colours; the powder sets into hard brittle laminae. These were placed in a box 6–8 in. broad and 3–5 ft. long; one end of the box was movable and could be pushed against the layered mixture. On removing one side of the box it was easy to follow the effects of pressure applied to the movable end of the box. It was found that not only were the layers folded, but increased pressure produced fractures; the hard material could not accommodate itself indefinitely to the strain and cracked across, as more pressure was applied the crumpled layers on one side of a crack were forced over those on the other side, overthrusts were produced and layers that were originally near the bottom of the box were pushed over layers that had been above them: the order of formation was reversed. This experiment gives a true picture of the effect of strains, compressions and tensions which have repeatedly changed the face of the earth's surface: hard and apparently unyielding rocks have been bent, folded, and broken. Under enormous pressure and high temperature rocks become plastic and bend: they also behave as a brittle substance and, as in the above experiment, become fissured and crack. Such fissures in the earth's crust are known as faults and many examples are shown on geological maps. Faults are by no means confined to rocks that have been folded; they occur also in horizontal rocks. The angle of inclination of the fault-plane varies considerably, many are almost vertical fractures but, as we shall see in a later chapter, some are more nearly horizontal than vertical. Coal miners are familiar with faults: after following a seam of coal for some hundreds of yards they find it suddenly comes to an end, its place being taken by another kind of rock: at the junction between the coal and the other rock there is much broken material made by the action of disruptive forces. What happened is this: the coal seam with the rocks above and below it was fractured under strain and the rocks on one side of the rent slipped down or were forced upwards so that continuity was broken. In regions where compression and crustal movement have been exceptionally great, blocks of rock are known to have been pushed several miles along an inclined plane of fracture over the torn surfaces of rocks that were originally above them.

Looking at a series of rocks exposed on a hill-side or in a valley one occasionally notices that some of the layers are more or less horizontal whereas other layers lying below them are inclined at a considerable angle. An example will make this clear. In Fig. 9, p. 225, the lower layers are more highly inclined than the rocks above them. The surface of the lower beds is uneven and, it may be, a conglomerate lies between the upper and the lower series of beds. This junction between the rocks where the boundary has been worn down into irregularities is called an unconformity. The lower, more steeply inclined rocks are part of an upraised sea-floor which became land; at a later stage the land subsided, some of it remained above water-level and some of it lay under a shallow sea where, at the base of the cliff formed by the higher portion, the waves piled up a shingle beach that is now a conglomerate. As the ground sank to a lower level sand and mud were deposited on its submerged surface. The important point is that the unconformity is evidence of a long interval during which there were oscillations in the level of land and sea.

In order to appreciate the extent of country covered with rocks of the same geological period and representing similar conditions of sedimentation we shall take a journey from the Dorset coast, starting from the white cliffs of Bat's Head, a few miles south-east of Dorchester. From Dorchester we shall travel in a north-easterly direction over Salisbury Plain to Marlborough, then eastwards to High Wycombe, through

Dunstable to Hitchin, past Cambridge, through part of Suffolk and Norfolk to the East Anglian coast on the borders of the Wash. In the country traversed there are many different scenes and yet there is a remarkable uniformity in the colour of the soil, in the curves of the higher ground and in the whiteness of the rocks exposed in pits and the scarred flanks of rounded hills with a sparse covering of soil supporting a characteristic vegetation in which beech plays a conspicuous part. Crossing the Wash and bearing to the north-west along the rising ground overlooking the flat seaboard of Lincolnshire, then across the Humber, still keeping to the same line of hills and bearing towards the east, we come to Flamborough Head, where on a grander scale than on the Dorset coast we end our journey as we began it, on a chalk cliff. We retrace our steps to Salisbury, and instead of going north, we turn east to Winchester and from there along the ridge of hills into Sussex and so on over the Downs to Lewes: bearing slightly towards the south we reach an abrupt terminus of the chalk at the majestic bastion of Beachy Head overlooking the English Channel, across which the English downland was once joined to the white cliffs of France. Returning to Winchester we follow the Downs to the north and a little south of Basingstoke turn eastwards to Dorking, past Rochester and south-east to Shakespeare's Cliff at Dover, where again we see the broken end of a great rib which at no very distant date united the land that is now an island to the land we know as France. In this expedition it would require more than passing glances at the scenery to convince a traveller of the persistence of the same kind of rock. Where the ground is relatively high it is easy to recognize the familiar qualities to which the downland owes its charm, but in many districts the chalk is hidden by gravel and other surface accumulations. These traverses illustrate the dominant rôle played by a particular kind of rock in controlling and moulding scenic features.

The next tour is selected as an example of the variety and range in structure and scenery which is readily appreciated as we travel across the breadth of England. Starting from Lancaster we follow the L.M.S. railway in an easterly direction to Giggleswick in Yorkshire. Lancaster is situated on rocks which are hard and compact masses of rather coarse sand known as Millstone Grit: these rocks form a ridge to the south-east of the town and the profile of some of the hills seen against the skyline ends abruptly in a shorter and steeper face at the end of a long and gentle slope dipping away from it. This characteristic feature is often seen in districts where relatively hard and soft rocks form an alternating series of beds with a slight inclination. A river will more easily cut a channel in the soft beds of shale, but when the harder rock is reached the valley will be broadened by the removal of the less resistant material; as that is eroded, the overlying harder rock will gradually fall away along lines of weakness due to joint-planes at right angles to the bedding. This is the process by which escarpments are made. At Giggleswick the scenery changes, bluish grey cliffs replace the more sombre and less impressive grits. We have passed from one subdivision of the Carboniferous formation—the Millstone Grit which was deposited in the delta of a river flowing from a northern land—to another and older stage of the same period represented by the Carboniferous Limestone which is made of shells and corals and a great variety of other marine refuse from the floor of a clear sea. From Giggleswick we pass along the valley of the Ribble with flat-topped Ingleborough on the left: turning east we make our way to Grassington, where the limestone hills give place to hills and escarpments of grit, a repetition of the surface features farther west. From Grassington we continue east over a belt of gritty rocks to Knaresborough where a different kind of rock that is rich in lime and magnesia lies as a band from north to south on a sloping foundation of Millstone

Grit. The Knaresborough rock is a limestone of the Permian period and the source of the so-called petrifying springs that should rather be called encrusting: it trends north and near Darlington bends eastwards to the coast, where it forms the buff-coloured cliffs from Hartlepool to South Shields. From Knaresborough we reach the Vale of York, which owes its flatness to the softer and less resistant nature of the underlying sedimentary beds deposited in inland lakes in a semi-desert country of the Triassic period. From York, where the actual surface is made of material deposited from an ice sheet on the Triassic platform, we continue north-eastwards to the steep face of a plateau where harder rocks of the Jurassic period form the western margin of the East Yorkshire moors bounded on the east by the sea. The Jurassic rocks of the undulating moorland and the long line of cliffs from Saltburn to beyond Scarborough include sandstones, grits and shales and some limestone, a series which can be traced through Lincolnshire along the cliff on which stands Lincoln Cathedral, through Rutland, Northamptonshire, Oxfordshire and finally to the coast at Portland Bill.

We will now make one more traverse: starting from Caernarvon Bay in North Wales we ascend Snowdon, one of many mountains in Wales and the English Lake District built in great part of volcanic ash and marine mud ejected and deposited on the floor of a sea in the Ordovician period long before the stage of earth-history represented by the limestone hills of West Yorkshire, Derbyshire and south-west England. Through Denbighshire to the Vale of Clwyd, along which the river Clwyd flows to the sea at Rhyl, we soon reach an entirely different set of rocks, some volcanic in origin, others that are upraised beds of sand and clay. Farther east we come to Carboniferous Limestone and Millstone Grit: then over the Cheshire plain where the red soil and occasional meres lie on relatively soft rocks containing salt beds of commercial value: these beds, like those under the plain of York, are of Triassic age; they were deposited in lakes comparable in the high percentage of salt with the Caspian and the Dead Sea. Farther east we are once more in a Carboniferous country on rocks that are continued north to Derbyshire and Yorkshire. Resting on inclined beds of Millstone Grit are sandstones and shales with intercalated seams of coal reached by the shafts of Nottinghamshire collieries. To the east of the coal-bearing series (Coal Measures) is a strip of limestone, the same we saw at Knaresborough, and beyond is lower ground made of red rocks like those of the Cheshire plain and the Vale of York. Passing into Lincolnshire we see a ridge of Jurassic rocks which is a direct continuation of the cliff from which the cathedral of Saint Hugh of Lincoln looks down on the flatland to the east, and the ridge of later date made of chalk, and beyond that the sea.

In travelling over the country as we have done we are constantly made aware of the close connexion between the structure and relative hardness of rocks and the varying scenic features. There is, for example, a marked contrast between the Snowdon range and the Pennine Chain, which are made of very different rocks, different in age, origin, and texture, in the type of scenery they produce. We have not as yet seen all the constituents of the rocky mosaic of England and Wales; we have not met with examples of the oldest geological period which are described in Chapter XVI, but enough has been said to give a general idea of what is meant by the geological structure of a country as seen in surface-view. From observations at the surface supplemented by data furnished by natural and artificial exposures, it is possible to construct geological sections which present a picture of the inner structure of the crust such as would be seen were it possible to cut a deep trench through the rocks from one side of the country to the other. In a horizontal section drawn from North Wales to the North Sea some of the layers of rock would appear in regular, parallel order; others would be less regular, inclined to one another at different angles and unconformable; some would appear crumpled and folded, and in many sections we should see breaks in continuity caused by fractures, or faults. All these features, revealed by sections, furnish valuable information from which it is possible to see in the disposition of the rocks an epitome of their history. One striking fact that emerges from a section drawn from Wales to the North Sea is the occurrence of older formations on the west and the younger rocks towards the east: the western rocks are relics of the most ancient portions of the crust on the flanks of which successively younger sedimentary beds were deposited.

In structure and scenery Scotland presents a striking contrast to England. Parts of the Highlands may be compared to the mountains of Wales, but on the whole the country north of the Border is a land *sui generis* both in the plan of its foundation-stones and in the evidence of stupendous folding and disruption recorded in their structure and arrangement. The following brief description is intended to serve as an introduction to references to Scottish geology in later chapters. In England and Wales rocks of the oldest periods of earth-history are rare and confined to comparatively small areas: in Scotland they play a much more important rôle both in their wider distribution and as dominating features in the scenery. To the great majority of people a view from one of the higher vantage points in the Highlands, such as the summit of Ben Nevis, would convey the impression of a jumbled collection of mountains and glens with no indication of an earlier and more uniform plan. If the same scene is viewed with broader vision and the inequalities are fitted into a comprehensive picture based on a truer perspective it is possible to superimpose upon the present confusion a quality of orderliness and to obtain a glimpse of an earlier stage in the development of the Highland landscape. Two geologists unrivalled in their knowledge of the geological structure of the Highlands and Lowlands wrote: 'Scotland may be regarded as a mountainous and hilly country, but, in reality, it is a dissected tableland of no great elevation and of extremely complex geological structure.' This is an illustration of the ability of a trained observer to reconstruct from apparently inextricable confusion the setting of the scene as it was. If, disregarding minor differences in height, we draw a line over the mountain tops and higher ridges, we realize that we are looking at a high tableland before it came under the destructive effects of erosion and denudation; we see a block of the earth's crust that in the course of millions of years was carved into its present state by rain, frost, and other tools acting upon hard and relatively soft material and aided by lines of weakness caused by fracture and dislocation. When we go deeper into the foundations of the Highlands the task of visualizing the past becomes much more difficult; as knowledge increased new problems emerged and many early interpretations of the geological structure had to be abandoned. The secrets of the Highlands have been partially but only partially unravelled: facts wrested from the rocks furnish the most remarkable example in the British Isles of the behaviour of the earth's crust under intense pressure both as a brittle and a semi-plastic material. It is in the Highlands of Scotland that the comparison with the behaviour of waves in a turbulent sea, based on a description by Conrad, is most applicable (see p. 8).

Scotland is naturally divisible into four major regions based on natural features.

I. The Northern and North-West Highlands, including Caithness with Duncansby Head at the north-eastern corner, Sutherland, Ross and Cromarty, and farther south to the south-western part of Inverness-shire. The southern boundary of the Northern Highlands is Glen More or the Great Glen stretching from Loch Linnhe in the south-west, the loch which starts from the eastern end of the Sound of Mull and extends far into the mainland, passing through Loch Ness to Inverness and the Moray Firth. This diagonal line marks the position of a fracture known as the Great Glen Fault, which divides the Northern from the Grampian Highlands, along which the once continuous layers of rock were dislocated and those on one side slipped sideways far beyond their former position. Within the North-West Highland region there are two clearly defined sub-regions; one of these, including the oldest rocks in Scotland, is a narrow strip more than 100 miles long on the edge of the western sea-board from Cape Wrath and Durness to the Point of Sleat, the end of the most southerly promontory of the island of Skye. The strip includes the most ancient pre-Cambrian gneiss, thick masses of grit and sandstone, known as the Torridonian formation from Loch Torridon, belonging to a much later stage in the long pre-Cambrian era, and upraised sediments from Cambrian and Ordovician seas. The second sub-region consists of an inconceivably complicated series of rocks, mainly crystalline schists that were originally beds of sediment, and extends eastwards to the coast with the Great Glen Fault at its southern boundary; it is separated from the other sub-region on the west coast by a north and south fracture known as the Moine thrust-plane, that is, a fault of low inclination along which enormous blocks of disrupted rocks were thrust a distance of many miles from their original position. The rocks on the eastern side of the Moine thrust-plane include the Moine Schists, believed to be pre-Cambrian, though their precise age is still under discussion, and large areas of Old Red Sandstone, which forms the high vertical cliffs at the north-east corner of Caithness and occurs in patches along the edge of the North Sea to the Moray Firth.

II. On the south-eastern side of the Great Glen are the Grampian Highlands, made in great part of Dalradian (named from the old Celtic region of Dalraida) schists but including a considerable variety of igneous rocks, which form the highest mountains in Britain. The Grampian, or Central, Highlands are bounded on the south by a great fracture parallel to the Great Glen, the Highland Boundary Fault which cuts diagonally across the country from the cliffs at Stonehaven a few miles south of Aberdeen in a south-westerly direction across the valleys of a series of rivers, the Esk, the Tay, and others flowing from the mountains on the north, through the southern end of Loch Katrine and Loch Lomond to Greenock on the Firth of Clyde. The northern border of the Grampian Highlands is the straight coast-line from the Moray Firth to Kinnaird Head, and the eastern boundary is the coast from Peterhead and Aberdeen to Stonehaven. In addition to the large areas occupied by schists like those on the eastern side of the Moine thrust-plane and by others more varied in composition there are enormously thick piles of Old Red Sandstone beds in both the Northern and Grampian Highlands regions, and it is certain that in former ages they were overspread by Old Red Sandstone as well as by younger rocks. The south-western border of the Grampian region is the fiord-indented coast immediately north of the Isle of Arran and including the Mull of Kintyre, Loch Fyne and the Firth of Lorne.

III. The third region, very different from the first and second, is called the Midland Valley or the Central Lowlands. It is limited on the north by the north-west–south-east Highland Boundary Fault and on the south by another roughly parallel fault separated by a distance of 50 miles; this is called the Southern Uplands Fault, and extends from Dunbar on the east coast by the Lammermuir Hills in a south-westerly direction to Loch Ryan across the Rhinns of Galloway. The Midland Valley includes the Firth of Tay, the Firth of Forth, Edinburgh, Lanark, and Glasgow, and the south-western border is on the shore of the Firth of Clyde. The region is one of many examples of a *rift valley*, that is, a block of country not necessarily a low-lying valley or chasm, but none the less structurally a valley because it lies between two parallel faults along which the rocks slipped downwards. In this rift valley the country is undulating and the scenery varied, short ranges of hills and smaller isolated bosses rising above the more low-lying lands: this contrast in elevation is a direct result of differential weathering of harder and softer rocks. The narrow belt of hilly ground parallel to the fault on the north includes the Sidlaw Hills north of the Firth of Tay, the Ochil Hills not far from Stirling, the Campsie Fells, and the Kilpatrick Hills a few miles north-west of Glasgow; all these hills consist largely of rocks of volcanic origin belonging to the Old Red Sandstone and Carboniferous periods. Small bosses and sheets of hard intrusive rocks stand out as prominent features, such as the rock on which stands Stirling Castle, the smaller hill crowned with the Wallace Memorial, Arthur's Seat, Salisbury Crags and the Castle Rock at Edinburgh, and other necks and plugs of old volcanoes, North Berwick Law, Dumbarton Rock on the Clyde, the Bass Rock and others. In the Carboniferous and Permian periods the Midland Valley was the scene of prolonged and widespread volcanic activity. The lower ground is underlain by Carboniferous and Upper Old Red Sandstone sedimentary beds which rest on folded rocks of older periods.

IV. The Scottish region beyond the English border is known as the Southern Uplands or simply as the South of Scotland; bounded on the north by the fault along the diagonal southern border of the Midland Valley, its southern limit is the Solway Firth and the Cheviot Hills, and a small district including Berwick is politically included in England. This Southern Upland region is a belt of high ground from the Irish Channel to the North Sea, through country of smooth, rounded hills covered with grass or heather, with Moffat in the centre; the north-eastern end is barely twenty miles broad, from Dunbar to Berwick. In the south-western district blocks of granitic rocks at Criffel on the edge of the Solway Firth, in the Loch Doon district and elsewhere give a more rugged and wilder aspect to the landscape. Most of the region is formed of highly folded marine sediments of the Ordovician and Silurian periods; a much smaller area consists of Carboniferous and Old Red Sandstone rocks and patches of Permian beds. Here too there is evidence of long-continued volcanic activity in the Old Red Sandstone, Carboniferous, and Permian periods.

The Shetland Islands consist partly of crystalline schists, perhaps of the same age as those forming the Grampian Mountains, and also of Old Red Sandstone. The Orkney group of islands is carved out of a plateau of Old Red Sandstone at the south-west promontory of which stands the Old Man of Hoy, a pillar 450 ft. high of flagstones resting on a plinth of igneous rock. (See Plate VIII.)

The Outer Hebrides, from Barra Head to the Butt of Lewis, are outposts of the oldest pre-Cambrian rocks worn into smoothed surfaces by the ice sheets of the Glacial period. The most striking scenic features of the majority of the Western Isles are the terraced hills of basaltic lavas and the high rugged peaks of the Cuillin Hills of Skye: these islands are dismembered relics of a vast plateau of igneous origin (see Chapter IX). Rocks of the Mesozoic era are very poorly represented in Scotland by patches in a few localities; it is, however, certain that Jurassic and Cretaceous beds were formerly much more widely distributed over the older rocks and have been almost completely removed by erosion and denudation.

CHAPTER VI

Medals of Creation

A good many years ago when authors were less self-conscious in the use of titles carrying an implication of orthodoxy, one of the early British geologists wrote a book about fossils which he called *Medals of Creation*. What is a fossil? The name implies something dug out of the ground (*fossum* = dug), but that is an incomplete definition. The disinterred body of a cat is not in the strict sense a fossil; it was buried by man, not by Nature. A popular misconception of fossils is that they are things turned into stone—petrifactions: many of the most valuable and informative fossils are petrifactions; they are unfortunately rare. The state of preservation covers a wide range. It is not uncommon to find on the surface of a bed of sandstone, as for example slabs of New Red Sandstone in Cheshire and the Midlands, impressions of footprints made by animals that had walked over the sand when it was soft, leaving no other record. One often sees various kinds of tracks on a beach made by animals crawling over or burrowing into the wet sand and impressions of the feet of birds: similar tracks and imprints occur on rocks, and in the light of actual specimens found at other localities it is often possible to identify the creatures which made them. Some fossils differ very little from living animals or plants: bones of elephants, hyaenas, and many other former inhabitants of Britain preserved in caves and old river gravels are exactly like fresh bones; trunks of trees in peat, blackened like bog-oak, retain the woody structure intact: these are true fossils. The carcase of a mammoth discovered in 1803 in frozen soil on the banks of the river Lena in Siberia in lat. 70° N. was almost as fresh as when alive and the flesh taken from cold storage was eaten by wolves. In 1901, after a peasant had related the discovery of a 'great hairy devil', members of an expedition organized by the Russian Academy of Science found other bodies of well-preserved mammoths, one of which had a fractured pelvis and right fore-leg, probably the cause of death while the beast was trying to extricate itself from the frozen ground.

The state of preservation of shells varies considerably; many, especially those in sedimentary rocks of the Tertiary era, may easily be mistaken for shells picked up on a modern beach: some shells have been hardened and altered by chemical change brought about by percolating water charged with mineral substances. Many fossils retain nothing either of the shell or body of the animal; they are merely casts. There are several kinds of casts: in a sandstone quarry, for example, in a coal-mining district, one often finds stems showing external markings, also the shape and size of the plant, but no trace of the actual substance remains. A piece of stem embedded in sand crumbles and falls to pieces, leaving a cavity or mould in the encasing sediment; on the surface of the cavity there is a clearly defined impression of the outer face of the original stem. The mould may afterwards be filled with sand carried by percolating water and this filling material makes a cast of the stem, though of the stem itself nothing remains. Similarly one finds fossil shells lacking the shell and represented by a cast in stone made of material introduced into the space left on the removal of the calcareous substance by the solvent action of water. Casts of this kind occur in Portland stone; the form of the marine creature is reproduced in hardened chalky mud that replaced the soft body; between this internal cast and the surrounding rock is a narrow space originally occupied by the shell.

A cast of a stem in a sandy or shaly rock is sometimes covered with a film of coaly material which is all that remains of the transformed outer part of the plant. The greater part of a twig or leaf embedded in sand or clay disintegrates and the plant substance is eventually transformed into a thin layer of coal on which the original surface-features may be clearly visible, such as the veins on a leaf or the pattern of the bark. In recent years methods of chemical treatment have been devised by means of which the black film is rendered transparent enough to be microscopically examined: in this way it is possible to make out the structure of the cells composing the superficial skin, or cuticle, of leaf or twig and thus obtain valuable and trustworthy evidence of relationship to existing plants. The great advance in our knowledge of the vegetation which grew on the borders of Old Red Sandstone lakes is mainly due to the successful use by Professor W. H. Lang of Manchester of an ingenious technique enabling him to extract from small and hopeless-looking scraps information of the greatest value. Fossils which seem worthless should not be discarded by collectors before they have been submitted to experts familiar with modern methods which often give almost magical results. One of the best examples of the possibility of obtaining much out of very little is the description by Professor T. M. Harris of Reading of a plant long known as *Naiadita*, which grew in a chalky ooze on the floor of a Rhaetic lake and occurs over a considerable area at many localities from east of the Mendips in Somerset to South Worcestershire and Warwickshire. The specimens, even the best of them, are in small pieces and at first sight furnish little or no evidence of their botanical nature. The name *Naiadita* was given to this fossil in 1844 because it was believed to be allied to the small submerged flowering plant *Naias*, that has an almost world-wide distribution. Many people collected material and made guesses on affinity, but no satisfactory conclusion was reached until Professor Harris, with the help of a simple technique, succeeded in converting speculation into well-established fact. The specimens—some of the best are from the Bristol district (Redland)—are usually very small bits of slender stems not exceeding 1¼ in. in length, with very thin leaves like minute blades of grass or sometimes narrowly oval. By immersing pieces of the rock in paraffin (kerosene) and focusing the microscope on to the fragments it was possible to see the structure of the cells and to detect well-preserved female reproductive organs, practically identical with those of mosses, small cups containing gemmae, that is, minute buds which are easily detachable and serve as a means of reproduction, spore-capsules with spores, and numerous tubular hair-like cells which served as hold-fast and absorbing organs. Some of the best spores were obtained by dissolving the lime in the rock in hydrochloric acid and then removing the sandy (siliceous) material by maceration in hydrofluoric acid. For a full and well-illustrated description reference should be made to Harris's *British Rhaetic Flora*, published by order of the Trustees of the British Museum in 1938. The important point is that these discoveries definitely settled the nature of *Naiadita* and proved beyond all doubt that it is a member of the class Bryophyta, which includes Mosses and Liverworts; the latter are rather simpler plants than mosses and commonly occur on damp gravel paths, walls, tree trunks and by hedgerows. *Naiadita* agrees most closely in general appearance with some of the smaller living mosses; in other respects it is nearer to liverworts, to which subdivision it is assigned. It is an extinct type though closely linked with existing genera of Bryophytes, a Rhaetic member of the ancient group of Liverworts which was represented by various forms in the forests of the Coal Age.

Fossils are occasionally preserved on the spot where the animals or plants lived: remains of coral reefs mark the site of reefs built up in vanished seas by coral-forming polyps (Chapter III, p. 40); masses of shells of the kind crowded together on a rock tell us of banks of oyster-like and other bivalves; stumps of trees with roots penetrating the rock are relics of forests preserved *in situ*. Remains of marine molluscs in a bed of shale, sandstone, or limestone are shells that fell to the bottom of the sea on the death of the animal which lived in them. In many instances fossils were preserved in sedimentary rocks far from their original home: pieces of petrified wood, seeds, fruits and twigs as well as bones and shells were carried long distances by currents, as drift-wood and other refuse from the land are transported many miles beyond the mouth of the Amazon, the Ganges, and other large rivers.

Reference has already been made to petrifactions, that is, fossils in which the internal structure has been rendered permanent by the infiltration of carbonate of lime or silica and the gradual replacement of the delicate framework by mineral substances. In order to examine microscopically a petrified stem or other plant fragment sections are cut by a revolving metal disk and then ground to a transparent thinness and mounted on a slide; another and simpler method is to treat the flattened surface of the specimen with acid which dissolves the petrifying mineral but leaves intact the plant tissues; a thin layer of a cellulose solution is spread over the fossil, and when peeled off it has embedded in it the plant cells and can be mounted for the microscope. Looking at a section or film under the microscope it requires an effort to realize that the delicate network of cells is part of a plant which lived more than 200 million years ago in a forest of the Coal Age. The cells are as distinct as those in a section cut from a living tree and we seem to see them not as lifeless things but as cells still playing their part as units in a living whole.

In this short and sketchy account it is impossible to mention more than a few of the many kinds of fossil animals and plants: for further information reference must be made to text-books on Palaeontology, that is, the Science of Ancient Life. In later chapters some account is given of fossils of different ages, and in the final chapter an attempt is made to convey a general idea of some of the numerous problems on which light is thrown by records of life discovered in the rocks. It must be remembered that the records available to us are but small in proportion to the numbers of animals and plants which left no memorial. As Darwin said, the fossils in the earth's crust are collections made at hazard and at long intervals: the geological record is necessarily very fragmentary and imperfect and gives us only a glimpse of the teeming life on the land and in the seas of former ages. The best way of forming a true estimate of the imperfection of the geological record is to observe what is happening now. With few exceptions rocks containing fossils were originally sediments on sea-floors, in lakes, river deltas and estuaries. Knowledge of plants of former ages is based mainly on twigs, seeds and leaves carried by rivers with sand and mud—samples of vegetation blown by wind on to the flowing water or contributed by plants dislodged by the undermining of banks. The important point is the restricted area from which specimens preserved as fossils have been derived: animals and plants living in places beyond the reach of streams and rivers would crumble to dust without leaving a legacy to future ages. Old surface-soils, with certain notable exceptions (see Chapter XIII), are rare; the great majority of fossil plants occur in beds of water-borne sediment deposited some distance away from the place where the forests grew. The records of plant life are of necessity not in the fullest sense representative samples and give only a partial and it may be a misleading picture of the contemporary vegetation as a whole.

Limestones are often rich in shells and corals and occasionally remains of seaweeds which had their delicate bodies encased in a protective calcareous coat: from such fossils it is possible to obtain a general impression of those inhabitants of seas which were protected by resistant, limy skeletons. The probability of preservation of land animals is small; skeletons of animals dying a natural death in their regular haunts would gradually decay and leave no memorial unless they happened to be within reach of flowing water. In some localities many large and complete skeletons of extinct animals have been buried in sediment within a small area in circumstances that were exceptional. Several years ago more than twenty complete skeletons of the extinct *Iguanodon*, a creature which walked on its hind legs like a kangaroo and was about 14 ft. high, were found in rocks of Cretaceous age (Chapter X) in a deep ravine that had been cut by a river in much older beds at Bernissart near Mons in Belgium: the occurrence of the skeletons crowded in a small space suggests that the animals were overwhelmed by a sudden flood of water. A bed of Old Red Sandstone in Scotland strewn with well-preserved fishes was aptly described by Hugh Miller as 'a platform of sudden death'. The Rancho La Brea asphalt deposits near Los Angeles, California, provided a varied assortment of bones of animals which had been caught in the sticky mass, many birds that alighted on the unsuspected death-traps and mammals lured to the spot in the hope of seizing helpless victims were themselves entangled. Looking at the pool of dark oily liquid in which large bubbles rise to the surface one can picture the tragedy and at the same time appreciate its value as a source of information to human students of extinct life. Other examples of the preservation of animals and plants are given in later chapters: enough has been said to give point to the statement that we can never hope to obtain sufficient information from fossils to enable us to reconstruct more than a very small proportion of the life of other days. It must be remembered that the recurrent crumplings and dislocation of the earth's crust are another contributory cause of the imperfection of the record. Rocks are Nature's palimpsests analogous to papyri on which scripts have been obliterated and overwritten by others.

Plants occasionally occur in rocks of igneous origin, especially in volcanic ash: stems and branches of trees in a forest smothered by showers of ash from a neighbouring centre of eruption have been obtained in a wonderful state of petrifaction from volcanic deposits of Carboniferous age in southern Scotland and the island of Arran. The conditions were much the same as we see to-day in the forests growing precariously on the slopes of Vesuvius and other active volcanoes. Vulcanicity is not only destructive; forces inimical to life have from time to time played their part in assisting us to a knowledge of plants which they almost completely but not wholly destroyed: some fragments remained, turned into stone by the infiltration of water charged with petrifying minerals derived from the volcanic ash. It is unfortunately rare to find flowers well enough preserved to be identified with confidence: some of the best examples occur in amber. Amber washed by the Baltic on to the shores of northern Germany from submerged Tertiary sediments is the slightly altered resin which exuded from wounds on the stems of pines and other trees and entrapped insects that had incautiously alighted on its sticky surface together with flowers and other plant fragments carried by wind to the resinous trickle.

Fossils are not only of great value as sources from which it is possible to gain some insight into the mysteries of evolution, they also serve as guides to the determination of the geological age of a rock; not only do they enable us to recognize the period to which a rock belongs but in some instances they give more precise information and enable us to date beds within narrow time-limits. In previous chapters two examples of the value of fossils as tests of age on a restricted scale were briefly described: species of sea-urchins in Chalk and species of graptolites in early Palaeozoic rocks. Other examples are mentioned in later chapters: the fossils of greatest help in fixing age within narrow limits are those of animals and plants with a short vertical range, species which are known to be examples of rapid evolution, one type changing into another in a comparatively short space of time in contrast to those that were more conservative and persisted with little modification from one period to another. In the accounts of geological periods attention is called to the range in time of some classes and groups of animals and plants, but no attempt is made to describe the distinguishing features of the various divisions of the two kingdoms: for further information readers are advised to consult text-books on Palaeontology. Although many fossils, even some from the older rocks, bear a fairly close resemblance to living animals and plants, the similarity is not enough to justify their reference to living species. The term 'species' is used for a collection of individuals resembling one another sufficiently to be designed by the same specific name: the precise significance to be attached to what is called a species has long been the subject of acute controversy. It is largely a subjective question, where to draw the line between one of Nature's types and another, all of which are merely stages in a continuous flux of form and structure. The term 'genus' is applied to a collection of species having certain features in common and believed to be descended from the same set of ancestors. As we pass from younger to older rocks the number of existing species rapidly decreases: in beds of Quaternary age most of the fossils belong to genera and species that are still with us, only a few are extinct. In the Tertiary era the more recent sedimentary beds have yielded a much higher percentage of existing species than the beds occupying a lower position in the geological sequence and therefore of greater age. Passing to still older rocks all the species and many of the genera are extinct, that is, they differ too widely from living animals and plants to be given the same specific name and in most instances the same generic name. It is of course very difficult to say when a presumably extinct animal or plant ceased to exist: no one believes that Dinosaurs are now at large anywhere in the world; they lived before the days when they could have had as associates men interested in the preservation of dwindling races. We can only speculate on the causes of extinction; but in all probability the 'Terrible Lizards' (Dinosaurs) of the Jurassic and Cretaceous periods were ill-equipped as competitors in the struggle for existence (Chapter xvII). It is known that some exceptionally interesting animals have been wiped out by the 'lords of Creation'. The Dodo, a bird about the size of a pigeon with wings inadequate for flight, was destroyed by early settlers in Mauritius; the Maoris were responsible for the extermination of the New Zealand Moa, a much larger bird also incapable of flight; Russian sailors in 1782 destroyed the last Sea Cows (*Sirenia*), aquatic animals related to elephants, on the shores of Behring Strait. The Great Auk, a bird comparable to the Antarctic Penguins, which formerly lived in the Shetland Islands, Iceland, and Greenland, became extinct as recently as the middle of the nineteenth century.

Traversing the ages we find many instances of both animals and plants characterized by unusually large dimensions, creatures that were gigantic as compared with their nearest living relatives; they seem to be examples of Nature's less successful experiments which flourished for a time and then with apparent suddenness died out never to reappear. Occasional reports from many sources of sea-serpents and Loch Ness Monsters have raised unfulfilled hopes of the survivals of animals believed to have long been extinct; but there remains the possibility of discovering in the little-known depths of the sea living descendants of creatures included in genera and families hitherto known only as fossils. A short time ago a fish 5 ft. long, metallic blue in colour, covered with large scales and distinguished by a broad and short tail, was caught in a trawl-net off the South African coast near East London. Unfortunately the putrefied soft parts were thrown away and only the skin was preserved. Examination by experts of what was left showed that the fish was very closely related to specimens of Jurassic and Cretaceous fossils included in a family believed to have been extinct for at least 80 million years. This solitary relic, named *Latimeria* after Miss Latimer, Curator of the East London Museum, is a member of a family of fishes which was previously known to have lived from the Devonian to the Cretaceous period, and now we know that one genus remains as an anachronism in the southern ocean.

Fossils help us in many ways towards a better understanding of the present as well as the past; they throw light, for example, on the present and often puzzling geographical distribution of plants and animals throughout the world; they also raise difficult problems connected with the climate of former periods. From the records of the rocks as from human records knowledge is gained which adds enormously to our enjoyment and appreciation of the present; we see the living creation in a new light, facts furnished by fossils enable us to distinguish between survivals from far-off ages and comparatively recent products of evolution, to grasp the significance and interpret the message of living animals and plants which speak to us as echoes from periods infinitely beyond the span of the human race.

CHAPTER VII

An Arctic Britain

This chapter gives a general account of the first of a series of journeys across the whole span of geological history. Contrary to the usual method adopted in books on the history of the earth, our retrospect follows a reversed chronological order, beginning with the most recent and passing downwards and backwards to the most ancient period. The stages covered in this chapter may be described as the last act of an age-long drama which covers more than half a million years and includes a brief and necessarily incomplete summary of events between the dawn of the prehistoric age, when primitive man hunted animals that disappeared long ago from Europe, and the historic period when man became the chief actor. The fragmentary pages of the story-book of the earth that have escaped the ravages of time are inscribed in the rocks beneath our feet, rocks lying for the most part in the order of their formation and relative age though not infrequently folded, crumpled and fractured and in some districts forced by irresistible strains and stresses into inextricable confusion. This solid foundation does not tell the whole story; a gap of several hundred thousand years separates the events chronicled in the rocky layers from those that belong to the latest and most recent chapter of geological history; the present or recent period in which we are now living was preceded by an age insignificant in duration as contrasted with the time covered by the earlier ages, and yet, as measured by human standards, too vast to be readily appreciated. Scenes from this comparatively modern age have been reconstructed not from the layers of solid rock but from various kinds of material scattered over the underlying firmer foundation and bearing no obvious relation to it.

Let us now briefly consider these geological accumulations. On an ordinary geological map the rocks of the solid foundation are distinguished by different colours showing their relative age and position in the geological table, but when we look at the surface of the country we notice many large areas where the rocks marked on the map are not actually exposed; they are hidden under moorland, woods, and cultivated ground, and it is only in cliffs by the sea, in hilly districts and occasionally protruding knolls and in deep river valleys that it is possible to recognize the rocks represented on the map as continuous bands or sheets. The Geological Survey publishes another kind of map on which the actual surface as we see it is shown: on it are marked the various superficial materials that form a partial and patchy mantle over the older rocky substratum. This superficial material is sometimes spoken of as drift or diluvium: the latter name is a survival from the time when the Flood was invoked as a convenient and all-embracing explanation of many natural phenomena. Under the comprehensive terms 'drift' and 'diluvium' are included boulder clays, moraines, deposits of gravel, sand, and mud in association with river channels, peat, and other surface accumulations. It is convenient to use the two technical terms, moraine and boulder clay; there are several other names employed by geologists for material closely related to moraines and boulder clays but these need not be introduced into a general and as far as possible popular account. Moraines are mounds or ridges made of a heterogeneous collection of pieces of rock mixed with gravel, sand, and clay; they form conspicuous features in the lower parts of many valleys in the Scottish Highlands, the English Lake District, in Welsh valleys, in some of the Yorkshire dales, and in lower ground far from hills as in the Vale of York and many other low-lying districts.

It is not surprising that the earlier geologists regarded the superficial material scattered far and wide over regions in the British Isles, northern Europe, and North America as evidence of a universal Noachian deluge. A staunch advocate of this orthodox view was Dr William Buckland, Professor of Geology at Oxford and sometime Dean of Westminster, whose book *Reliquiae Diluvianae*, published in 1823, is an interesting example of accurate observation and incorrect deduction. Fourteen years later a Swiss geologist, Louis Agassiz, shocked his scientific colleagues by attributing the supposed diluvial deposits to an entirely different cause: researches in Switzerland had convinced him that the mounds and ridges of moraines and other deposits associated with glaciers were the result of ice-action. His knowledge of the present led him to an interpretation of the past. He realized that the heterogeneous material in places beyond the reach of existing glaciers was proof of the former greater extension of ice. Buckland's desire to discover the truth led him to visit Switzerland and see for himself the evidence cited by Agassiz in support of his heretical conclusion. The two geologists then made a tour in Scotland, England and Wales. Agassiz, thoroughly familiar with many aspects of glacial conditions in the Swiss Alps, at once recognized many of the deposits in a country without glaciers as the exact counterparts of those in Switzerland. Buckland, true to the best traditions of a student of Nature who realizes that acknowledgment of mistakes is a virtue, admitted the force of his companion's contention and became an ardent supporter of the Glacial theory. After prolonged controversy and obstinate disinclination to abandon cherished beliefs, the incorrect implication of the descriptive name Diluvium, which is still occasionally used, was generally acknowledged: a new line of geological enquiry was thus inaugurated and, while there has been agreement in accepting moraines and other superficial deposits as proof of ice-action, there are still differences of opinion on the sequence of events in that phase of earth-history known as the Glacial period. The revolutionary change of view tended to foster a more scientific attitude towards the biblical story of the Flood. In 1890 the late Professor Huxley, after demolishing the old idea of a universal flood, wrote: 'The story of the Noachian deluge has no more claim to credit than has that of Deucalion; and whether it was, or was not, suggested by the familiar acquaintance of its originators with the effects of unusually great overflows of the Tigris and Euphrates, it is utterly devoid of historical truth.' He went on to speak of the story of the Flood in Genesis as merely 'a bowdlerised version of one of the oldest pieces of purely fictitious literature extant'. No one believes in a flood that was universal, nor that any natural phenomena in Europe are attributable to the deluge. Excavations at Ur in Mesopotamia in 1929 by Sir Leonard Woolley brought to light records of a flood that must have been on an exceptional scale: it was natural that the Assyrian writings should have described the great overflow of the waters as universal, that is, over the whole world known to them. As Sir Frederick Kenyon says in his book, *The Bible and Archaeology*: 'One cannot doubt that there must be some foundation, in fact, for a tradition which had fixed itself so deeply in the national consciousness; and here Archaeology, with its other hand (or rather, spade), has revealed physical facts in the site of the ancient city of Ur which furnish material confirmation of it.' This digression on the Flood, though irrelevant, is perhaps excusable as an illustration of the importance of preserving a sense of proportion in matters scientific as in all human affairs.

Another and more impressive proof of widespread ice-action

in Britain is furnished by a different kind of material, the boulder clay. This cannot be closely matched in Switzerland; it usually consists of a mass of stiff tenacious clay with innumerable embedded pieces of rock of many kinds, varying from stones a few inches in size to blocks over 100 yards in length. A characteristic feature of boulder clay is the general absence of a layered arrangement such as one would expect if it had been deposited from flood-water; the stones, scattered irregularly through the clay, are described technically as boulders or erratic blocks, the term 'erratic' having reference to the fact that many of the rock samples have been brought far from their original homes. Another striking feature is readily seen on examination of the shape and surface of the boulders; they are not completely rounded like ordinary water-worn pebbles but partially rounded; the edges are not sharp and the faces have been worn down into smooth surfaces on which can be detected series of scratches or striae lying more or less parallel one to another. Moreover the striated faces are smooth as if polished. Boulder clay occurs in many parts of Scotland, England, and Wales, not only in valleys but also on high ground and as a capping to cliffs in many coastal districts. It is spread over hill and valley alike and is often a conspicuous feature on plains. One of my earliest recollections of the days when Geology first interested me is a cliff of reddish boulder clay rising from the beach on the eastern edge of Morecambe Bay, at Hest Bank, a few miles north of Lancaster: the largest boulders are blocks of a light grey limestone exactly like the rocks of neighbouring hills; there are many other boulders which cannot be matched with rocks near at hand. Nearly all of them are scored with scratches and have smooth faces. Similar exposures of boulder clay occur in many parts of the country and everywhere they show similar features—a miscellaneous assemblage of scratched and partially rounded stones, some apparently identical with the solid rock in or near the locality, others that came from more distant sources. The question is, how was boulder clay formed? It has already been said that it differs from moraines associated with glaciers in Switzerland and other countries; the main difference is the smoothed and striated faces of practically all boulders. Is there anything in other countries that gives a clue? In Switzerland the valleys are occupied by glaciers carrying on their surface rows of rock-debris derived from overhanging cliffs, and these, with sand, gravel, and clay carried in suspension by streams issuing from the ends of glaciers, are deposited as rubbish heaps and layers of detritus in the lower parts of the valleys. In Arctic lands, such as Greenland and Spitsbergen, vast areas are covered with ice and for hundreds of miles a monotonous white mantle is spread over the country. Greenland is buried under an enormous sheet of ice except along a narrow strip of coast where great tongues of ice are thrust downwards and outwards as glaciers by the inland sheet into fiords and valleys. Looking at the cliffs of ice towering above the sea-beach or at the ends of fiords one sees bands of rock-debris and boulders embedded in the lower part of the icy walls. Another important fact is the occurrence, in close association with the ends of glaciers in valleys near the edge of the land, of masses of material precisely similar to boulder clay. Without going into further detail let us see what we can learn from the present conditions in Greenland and other Arctic regions. Ice overrides most of the mountains and is not confined to valleys: here and there near the edge of the ice-sheet a few jagged summits of the higher peaks rise like islands in a white sea, but for the most part there is nothing visible but ice. We have all of us looked from high ground in the English Lake District upon rolling cumulus clouds filling the valleys far up the sides of adjacent hills, leaving only the highest ground protruding through a sea of mist. Such a scene helps us to visualize the landscape as it was when, substituting ice for cloud, the greater part of Britain was smothered in ice.

But how do we know that there were ice-sheets over the British Isles? We know that over Greenland there is now a sheet of ice many hundred feet in thickness; we also know that except close to the coast the ice carries no stones on its surface because most of the rocks are not exposed. Impelled by pressure from the higher levels the enormous sheet moves with irresistible force over the ground, picking up loose blocks lying in its path. It has been calculated that a glacier 1000 ft. thick exerts a pressure of 25 tons on each square foot. The litter of rocks collected by glaciers and ice-sheets and wedged into their lower surfaces may be compared with the projecting pieces of metal on a rasp or file which make the tool an efficient instrument. The boulders are pressed against the rocky ground, scoring and grooving it, and themselves scratched and ground in the process. It is these boulders that furnish a clue to the origin and history of the erratics in British boulder clays. On the Greenland ice-sheet travellers have noticed rivers pouring through fissures into the depths below; such streams flowing underneath the ice carry with them the finer products of attrition such as gravel, sand, and clay. The mass of clay in which the boulders are embedded is made of the finer particles ground by the cutting-tools from the rocks below. Thus both coarse and fine materials accumulate under the ice and it is these which are left as boulder clay and sediment when the ice-sheet melts. Boulder clay, unlike moraines, was formed below moving ice and not carried on its surface.

We will now look more closely at boulder clays in this country and at the exposed surface of rocks in their neighbourhood. Let us first examine the surface of rocks in valleys, on the higher slopes, and on mountain summits. Walking through Scottish glens or valleys in the English Lake District, in the

Yorkshire dales and elsewhere, one can hardly fail to notice that exposed hummocks or knolls have a smoothed and rounded appearance as if they had been rubbed down by some giant abrading tool (Plate III); a closer inspection of the rocks often reveals grooves and scratches. It may be said that scratches mean little and may be due to many causes; but it will be noticed that they usually follow a definite direction, they may be parallel to the direction of the valley though by no means invariably, and sometimes the trend of the striation will be found to be transverse to the valley. On some rocks that have a rounded form suggesting the passage over them of a glacier or ice-sheet it is not always possible to detect any striae: this is because long exposure to the weather has obliterated them, but, if one removes the covering of turf from a rock, series of scratches and grooves often become apparent. When hummocks of rock have a gently curved, rounded form and are grooved it is practically certain that they were once overridden by ice studded with stones that served as graving tools. Applying this, it is possible to say in many instances that ice passed over the summits of mountains 2000 ft. or more in height. Evidence of ice-action is also furnished by the surface-features of hills; the broadly sloping contours of Saddleback near Keswick at once recall rocks moulded by an ice-sheet. On the other hand, the jagged peaks of some mountains where no striated platforms can be seen owe their present outlines to the action of rain and frost.

Impressive evidence of the former presence of sheets of slowly moving ice is also furnished by the erratic blocks themselves, both those still enclosed in boulder clay and others scattered over the ground. A few examples will serve as illustrations of this important and interesting line of enquiry. Many boulders can be referred without difficulty by observant amateurs to the parent source, e.g. the light grey blocks of limestone in the boulder clay of North Lancashire on the shore of Morecambe

PLATE III

Glaciated pavement in Cumberland

Bay, which it is easy to see must have come from the neigh-bouring hills. There are, however, many boulders mixed with those of local origin that are foreign to the district, and here expert knowledge is needed. A geologist familiar with the rocks of different parts of the country may be able to name the source of origin at sight, but it is often necessary to examine microscopically thin transparent sections. One of the best examples of boulders which have thrown light upon the course taken by ice-sheets is furnished by the well-known granite of Shap in Westmorland, which forms an irregular oval mass about 2 miles broad from east to west and a mile from north to south. This rock is generally reddish or flesh-coloured and is distinguished from many other granites by the large pink felspar crystals, often an inch or more in length. Polished slabs and pillars of Shap granite may be seen in many buildings in London and elsewhere. The important point is that boulders of Shap granite can readily be recognized even by a layman, and their occurrence in boulder clay throws light on the direction taken by ice-sheets and glaciers, that is assuming there is no means, other than ice, by which large boulders could be trans-ported over long distances. Large and small erratics of Shap granite have been traced east of the parent rock into the Vale of Eden, and from there farther east across Stainmore Pass which cuts through the Pennine Chain at a height of about 1400 ft. above sea-level: the ice followed approximately the route of the Roman Road. On the eastern side of the Pennines the same granite occurs in boulder clay in the Vale of York; near Doncaster there is a large patch of clay containing striated boulders of Shap granite and rocks from the English Lake District. The trail continues to Darlington, thence to the York-shire coast at Robin Hood's Bay and several other places, also farther south in Lincolnshire. At Dimlington in East Yorkshire near Spurn Point a boulder is recorded measuring 3 by 1½ ft.

Another set of boulders has been traced south of Shap to Lancashire, Cheshire, and much farther afield. What then are we to infer from these numerous pieces of Shap granite lying about the country as though flung down by giants engaged in a rock-chase? Transport cannot be attributed to local glaciers: the fact that the boulders are strewn in abundance over Stainmore Pass is convincing evidence of the former existence of an ice-sheet large enough and thick enough to surmount the Pennine Hills. This raises another point: why was the ice not confined to the southern route which we know from boulders in Lancashire and Cheshire was taken by some of the ice that passed over the fells of Shap? There was another ice-sheet which came from the south-western part of Scotland, and as it travelled towards the south along the western edge of England and over the floor of the Irish Sea it came into contact with glaciers that had coalesced from valley to valley to form continuous sheets over the higher ground of Lakeland. The Irish Sea ice passed over the summit of Snaefell (2034 ft.), the highest ground in the Isle of Man, and occupied the trough that is now filled with water. It was the pressure exerted by the Irish Sea ice-sheet on another sheet moving along a more westerly route over Shap Fells, where it picked up loose blocks and plucked great pieces from the solid rock, which deflected part of the frozen flood towards the east over Stainmore Pass into Teesdale and farther east to the Cleveland Hills which were not completely overridden. We know from the distribution of other kinds of boulders that the ice-sheet which occupied the Irish Sea came from Scotland: boulders of a granite differing from that of Shap and without doubt identical with that which occurs *in situ* at Criffel in Galloway have been found in South Lancashire, in Cheshire, and as far as Birmingham in the Midlands. The occurrence of Criffel boulders with Shap granite on Stainmore Pass and the east coast shows that the Scottish ice-sheet followed the main routes, south and then south-east and from the Vale of Eden across the Pennines into Yorkshire. There are many other kinds of boulders which mark the routes taken by ice-sheets; boulders of old lavas from Borrowdale and other rocks from the Lake District, also boulders of a crystalline rock peculiar to Ailsa Craig, the island peak that rises above the sea in lonely grandeur at the entrance to the Firth of Clyde south of Arran. Erratics from Ailsa Craig have been found in boulder clay along the eastern seaboard of Ireland from the north to County Cork in the south, in the Isle of Man, in Anglesey and North Wales and on the shores of Cardigan Bay.

The distribution of boulders that came from parent rocks in Lakeland and other evidence shows that nearly all the high ground was covered with ice which left traces on Helvellyn up to 2800 ft. above sea-level and on Scafell as high as 2500 ft.: Black Combe (1969 ft.) was completely hidden. Some of the Lake District ice moved south into Morecambe Bay and far beyond. The hills of North Wales contributed other ice-sheets and glaciers. There is good evidence that the Welsh ice reached St David's Head but did not penetrate quite as far as the Bristol Channel; the southern limit of the ice is believed to have been in the neighbourhood of Swansea. From North Wales ice travelled eastwards over the Shropshire-Cheshire plain, north to the Irish Sea, west to Cardigan Bay and St George's Channel. Ice in the bed of the Irish Sea and the sheet from West Scotland impinged upon the glaciers that had their home in Wales. As Dr W. B. Wright says: 'The whole wild and rugged scenery of North Wales is in very large measure due to this glaciation of the not very distant past.' Erratic blocks from the Southern Uplands of Scotland have been found on the Cardigan coast and in Pembrokeshire. Nearly the whole of Ireland was covered; a series of moraines marks the southern limit of the ice in the Wexford district, round the northern flank of the Wicklow

Hills and the southern outskirts of Dublin to the mouth of the Shannon and the coast of Clare. In many of these glacial deposits boulders from Ailsa Craig are a noteworthy feature. There were glaciers in Kerry, but some of the southern districts were never completely submerged.

With the ordinary glacial deposits there are occasionally inter-mixed patches of sand containing marine shells which it is believed became entangled in the ice as it pushed its way over the floor of the Irish Sea. A sample of the bed of the Irish Sea has long been known at Moel Tryfaen between Snowdon and Caernarvon at more than 1300 ft. above sea-level: its occurrence raised much speculation; some geologists attributed the deposit to floating icebergs, a theory which assumes a submergence of more than 1300 ft. unsupported by any satisfactory evidence; a more probable explanation is that the material from the sea-floor has been carried to its present positions by ice moving over land.

The Cheviot Hills were partially and in places completely overridden. In the Highlands of Scotland sheets of ice spread over the greater part of the country, leaving ridges and mounds of boulder clay over the Lowlands. The Southern Uplands of Scotland were another centre from which coalesced glaciers formed ice-sheets that carried boulders northwards into the Midland Valley and came into contact with ice travelling south from the Grampian Highlands. Other ice from the Southern Uplands and the Grampians moved east to the North Sea, where it met an ice-sheet from Scandinavia. The undulating Lowlands belt known as the Midland Valley is strewn with glacial deposits; it lies across Scotland as an old rift-valley between the Highlands and the Southern Uplands. The Orkney and Shetland Islands were overridden by ice; the Inner Hebrides were almost entirely submerged, but the higher mountains of Skye and Mull nourished local ice-caps. Rocks of the Outer Hebrides bear the unmistakable impress of moving ice in their rounded, polished, and striated surfaces, except the highest ground in the northern part of Harris, which protruded above the sea of ice. Beyond the Outer Hebrides the ice-sheet continued its course some distance into the Atlantic Ocean where it floated as a precipitous wall like the Ross Barrier in the Antarctic.

So far attention has been directed mainly to boulders as sources of information from which to visualize the British Isles as they were when the Glacial period was at its maximum: there are many other kinds of evidence which must be taken into account. A very helpful guide is furnished by observing the general trend of grooves and striae on glaciated rocks in valleys and on hills: it is often noticeable that the striae over a wide area lie towards the same points of the compass, and this together with clues given by the distribution of boulders provides valuable data. The transport of boulders from Scotland over western England and Wales, boulders of Shap granite littered over a wide area from Westmorland to the Lancashire and Yorkshire coasts; these and many other facts contributed by enthusiastic collectors, several of whom took up geology as a hobby, afford proof of the severity of the climatic conditions in Britain when it was in very much the same state as Greenland is to-day.

There is, however, a still more impressive piece of evidence which enables us more fully to appreciate the contrast between the geologically recent past and the present. Among boulders collected at many places on the east coast of England were some which could not be matched with any British rocks. Fortunately it was not difficult for experts to identify some of the foreign erratics as larvikite, a very distinctive crystalline rock from Larvik near Oslo in Norway. One can often recognize polished slabs of this Scandinavian rock on the facades of buildings in London and many other towns by its blue colour and the iridescence of the large felspar crystals. A geologist gives the dimensions of a larvikite erratic found on a Durham beach as 5 by 3 ft.; he described the boulders as serving the double purpose of a facing to public houses and very useful indicators of the direction taken by an invading ice-sheet. An ice-sheet from Norway extended in a south-westerly direction over the bed of the North Sea and impinged upon the English coast, where its presence is proved by widely scattered erratic blocks, some of which have been found at inland localities, e.g. in Cambridgeshire and in the basin of the Upper Thames near Oxford. Assuming that the Norwegian mountains were not much higher in the Glacial period than they are now (8400 ft.) and bearing in mind that Norway is 400 miles from Scotland, some geologists have raised the question whether the urge of gravity acting upon an ice-sheet would be adequate as a propelling force; they suggest floating icebergs as an alternative. The more widely held view is that the Scandinavian erratics were transported by ice-sheets, and it may well be that the Norwegian Highlands formerly rose to greater heights.

The greater part of the British Isles was overwhelmed by glaciers and sheets of ice overriding hills, valleys, and plains, except in certain districts which bear no trace of glacial deposits and no grooved and scored rock-surfaces. The belt of country south of a line joining the Severn and Thames estuaries was beyond the ice-front. At this stage it is worth while to raise a point connected with the great Ice Age which has long been a controversial question: it is agreed that the Ice Age was of long duration, probably not less than half a million years; it is also agreed that there were oscillations of climate over a wide range of temperature. There was a phase of intense glaciation when the ice-mantle reached its maximum; there were periods when ice was confined to valleys in hilly districts as in Switzerland to-day; and there were other periods when the climate

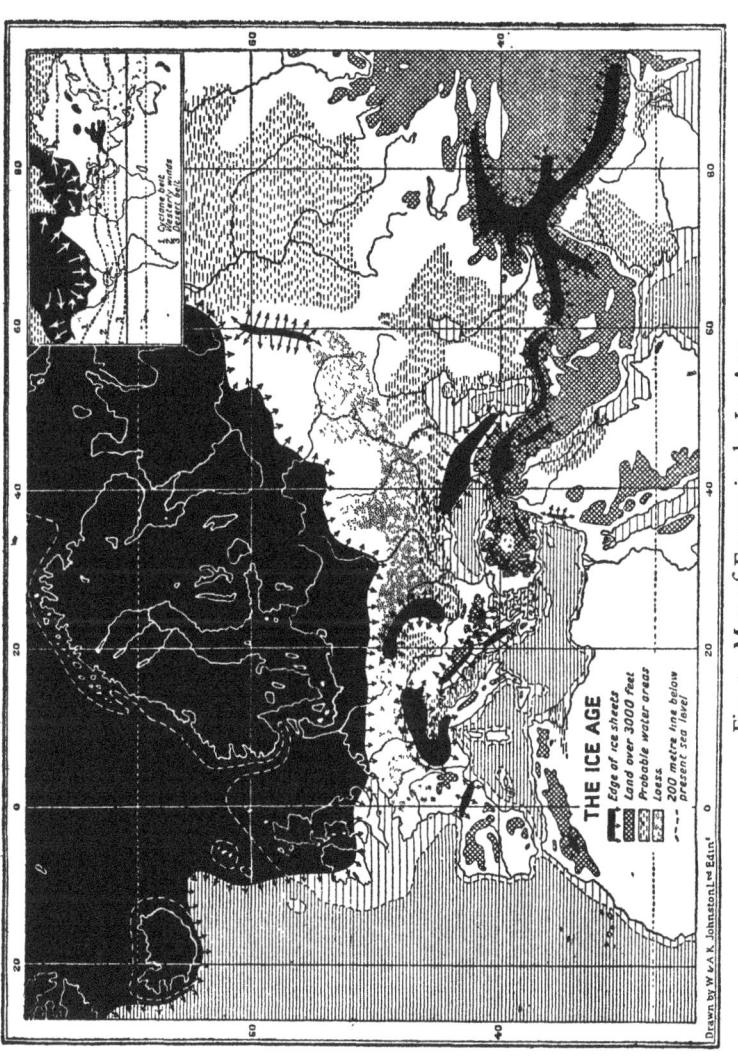

Fig. 4. Map of Europe in the Ice Age.

(After Professor Myres, from the *Cambridge Ancient History*, vol. I.)

was at least as genial as it is now, when for some thousands of years there was practically no ice. There were what are called glacial phases varying in severity and one or more interglacial, mild phases. The number of alternations within the Ice Age as a whole of arctic and temperate sub-periods is still *sub judice*. A deposit known as the Hötting Breccia near Innsbrück is a good example of an interglacial record: it rests on boulder clay and other glacial beds lie above it. In this breccia were found some plant remains, including a *Rhododendron, R. ponticum*, which now grows 5° south of Innsbrück, also a southern species of Box and a *Rhamnus* (Buckthorn) very similar to a Canary Islands species. It is clear that when these plants were living the climate could not have been glacial. It has been fairly well established that in the Swiss Alps there were four sub-periods of more intense glaciation and three interglacial phases; but it has not so far been possible definitely to establish a similar sequence in Britain. All that need be said is that we know, from remains of animals, including man, and plants associated with glacial deposits included within the time-limits of the Ice Age, that there were well-marked and considerable fluctuations in climate.

In the light of the conclusion stated above we shall now briefly consider some other proofs of the former presence of glaciers and ice-sheets in Britain. There are several districts in Scotland, England, and Wales where at some stage or other of the Ice Age fresh-water lakes occupied sites where there are now only river valleys or dry land: the lakes were caused by damming of valleys by boulder clay, moraines, or barriers of solid ice. One such extinct lake occupied the Vale of Pickering in East Yorkshire; it was about 25 miles long from west to east and 10 miles broad. The river Derwent now flows over the old lake-floor; the ruins of Kirkham Abbey, visible from the train as one travels southward to York, are near the old southern shore. Evidence on which this statement is made includes the occurrence of terraces of water-borne gravel and sand where the edge of the lake lay, and the discovery of delta deposits showing where rivers from the neighbouring Cleveland Hills, which were not covered by ice, discharged their loads of sediment. The eastern outlet of the Vale was blocked by a barrier of boulder clay and moraines on the margin of the Scandinavian ice-sheet as it forced its leisurely way over the bed of the North Sea. There was another lake near Shrewsbury; others in the Welsh Borderland, in the Yorkshire dales, and many other districts. The most famous of all lakes that no longer exist occupied Glen Roy to the north-east of Ben Nevis. On the sloping hill-sides curving downwards to the floor of the glen there is a series of three terraces giving at a distance the impression of roads, one above the other and following parallel horizontal courses. For many years these parallel roads of Glen Roy taxed the ingenuity of geologists and appealed to the imagination of Highlanders. They are natural shelves or terraces marking successive shore-lines of a lake that occupied the glen and at intervals shrank in volume: records of the dwindling were left at successive levels. The damming of the glen was caused by ice.

Another natural feature left as a legacy by vanished glaciers is the hanging valley: in Glen Nevis and many other Scottish glens, the Nant Francon Pass in North Wales, in Borrowdale and elsewhere in Lakeland one often sees lateral valleys between hills bordering the main depression which do not reach the bottom of the valley but are, as it were, suspended, and their mouths overhang the major valley. The lateral and main valleys are disconnected; water issuing from the former falls as cascades instead of along a continuous slope to join the stream below. The Lodore Falls near the southern end of Derwentwater are caused by water from a hanging valley. Explanation of these valleys is simple if we picture the main valley occupied and eroded by a thick glacier reaching far above the present ground-level where its surface would be as high as the ends of the tributary valleys that are now left hanging.

Glaciers are powerful agents of rock-destruction; they are forced down from the gathering grounds that feed them from above; their mass presses on the valley floor and sides, and the rocks incorporated into the lower surface of the moving ice act as efficient rasps and grinding tools; thus the valley is deepened, widened, grooved, and scratched. The result is different from that produced by a river: a river is also an efficient cutting-tool, but it cuts mainly into the middle of the bed and, unlike a glacier, has little effect, other than undermining, upon the sides of the valley; it makes a V-shaped valley in contrast to the broader and more rounded U-shaped valley characteristic of ice-action. A typical example of a U-shaped valley is seen along the road passing over Dunmail Raise between Grasmere and Keswick: looking along the valley towards the north one can hardly fail to notice the broadly rounded curves flanking the summit of the hill that overlooks the descent to Keswick. A glacier once passed along the course of the present road, its sides reaching far up the shoulder of Helvellyn. There are many U-shaped valleys in Scotland and Wales: Glen Rosa in the island of Arran is a good example; the sides of the valley rise in low curves above the stream below. The contrast between the smoothed and striated rocks in the lower slopes of a valley and the more jagged rocks at higher levels which the ice failed to reach is often clearly marked. In a valley that was formerly occupied by ice it is not only the grooves and scratches that remain as records; the surface of the floor is made irregular by hummocks and knells of rock, and these were rounded and smoothed by the grinding tools embedded in the glacier into curved rounded contours compared with the backs of sheep lying on the ground; hence the name *roches moutonnées*.[1] Streams of water issuing from the end of a glacier carry away the finer detritus produced by the wear and tear of the rocks, and this material becomes mixed with the boulders and moraine debris at the melting end of the ice or spread as sheets over the lower reaches of the valley.

Among the many convincing proofs of the former glaciation of a district is the occurrence of what are known as perched blocks: one example will suffice as an illustration. Walking along the Crummack valley about one mile north of Austwick in the West Riding of Yorkshire one sees two readily distinguishable kinds of rock; on the lower ground the grassy surface is interrupted in several places by exposed patches of a dark grit; the higher ground of Norber on the opposite side is capped with regular layers of white limestone. Scattered over the hillside are several blocks of dark grit each resting on a plinth of white limestone about 18 in. high. The dark brown blocks undoubtedly came from the rock *in situ* below; they are not rounded like boulders nor do they show any signs of ice-action: the surface of the raised white platform on which each piece of grit is perched is smoothed and here and there one can detect grooves and striae (see Plate II, lower figure). These perched blocks reached their present position when ice filled the valley; they fell on to the glacier and were carried away from the parent source: as the ice melted they were left stranded on the limestone over which the glacier passed. It is interesting to note that the limestone plinths stand up a foot or more above the general level of the ground; this is because the perched blocks of grit protected the limestone platform below them from the solvent action of rain. The perched blocks have been resting where we see them for several thousand years.

1 Moutonnée really refers to the knobby appearance of the tufts of hair in barristers' wigs.

Enough has been said to justify the conclusion that at one time, or more probably at different times in the course of the long Ice Age, by far the greater part of the British Isles together with the Irish Sea and part of the North Sea lay under sheets of ice. As previously pointed out, the ice-cover was by no means complete. Portions of the Pennine Chain and the Cheviot Hills were left bare. The Cleveland Hills in the North Riding of Yorkshire, though encircled, were not covered: some of the higher mountains in the Lake District stood out as islands of exposed rock. The summit of Cross Fell (2930 ft.) on the eastern side of the Eden valley probably remained nearly 700 ft. above the surface of the ice-sheet. In southern England there were caps of snow on the Cotswold and Mendip Hills, but they were beyond the ice-invaded territory; the Bristol and Gloucester districts were partially ice-free. Erratic blocks from West Scotland have been found in North Devon, but they are believed to have been dropped from floating icebergs detached from the great terminal wall of the ice-sheet that was not far off. Similarly, far-travelled boulders are recorded from Selsey Bill, a few miles south of Chichester in Sussex. We can in imagination see icebergs being carved from the edge of the ice to the north of the south coast, and as they stranded in shallow water they dropped their load of stones.

What was the state of the southern belt beyond the Severn-Thames boundary of the ice? There were fans of sediment, chiefly gravel and sand, spread by water issuing from below the ice. On the ground, bare of trees, were clumps of stunted herbage. It was bleaker and more barren than the moorland on British uplands or the Breckland of East Anglia; inhospitable and desolate; comparable with the tundra of North Europe and Siberia where the soil is permanently frozen to a few feet below the surface. Two other questions at once spring to the mind: was it possible for flowering plants to live in this tundra-like belt, and was it inhabited by man and other members of the animal kingdom? These questions can be only partially answered. Let us take man first: unfortunately comparatively few human bones and skeletons have been found in deposits left by the Ice Age. Evidence of human occupation is derived almost solely from flint implements. It has now been established that man lived in Britain during and before the Ice Age: we can hardly believe that he could have lived in regions where the arctic conditions were most severe, but life may well have been possible on the southern ice-free belt. Little is known of the men who left their crudely fashioned implements in caves and river terraces contemporary with the Glacial period; all that can be said is that men wandered in search of food over the hills and moors when the valleys were filled with ice. To-day in Greenland many Eskimo and a few Danish officials live on the west coast from near Cape Farewell to a comparatively short distance from the northernmost end: there are a few settlements on the less hospitable eastern seaboard. Similarly there is no reason why some of the ice-free districts of Britain, even during the colder spells of the Ice Age, should not have been habitable. Much has been written on the early history of prehistoric man; his gradual development as a maker of tools to supply his needs has been traced through a series of stages to which names have been given, chosen for the most part from places where implements and other remains have been found. The history of the human race is outside the scope of this book, but there are a few questions which cannot be entirely ignored. Many years have passed since the first flint implement was discovered near Gray's Inn Road, London, in 1690, nearly two centuries before the study of early man and his work became one of the most fruitful sources of controversy and the basis of a new branch of scientific enquiry. No satisfactory answer has yet been given to the question—at what precise stage in geological history did man make his appearance as a fully developed human being? We know that he was contemporary in England with cave bears, hyaenas, hippopotamus, rhinoceros, and more than one kind of elephant. Dependent for his food upon wild nature he trapped and killed animals, his home was a cave or a shelter built of skins and boughs. He was a primitive and brutish creature whose hands were able to convert shapeless stones into crude instruments for the chase or domestic use; he was able to give expression to an innate sense of the supernatural by carving on bone outline pictures of game and hunters, probably in the belief that they served as magic aids to success in hunting forays.

One of the few skulls of early Stone Age man was discovered in 1911 in a terrace of gravel on the banks of the river Ouse at Piltdown near Lewes in Sussex. This exceptionally important discovery, made by a lawyer, Mr Charles Dawson, a keen amateur geologist, stimulated the ingenuity and argumentative skill of experts in many lands. Unfortunately, when the workmen in the gravel pit first found the skull they took it for a coco-nut and treated it accordingly until its value was recognized by Mr Dawson. Subsequently a lower jaw was found, and then followed an exciting debate on the all-important question—was the jaw part of the Piltdown skull or did it belong to another skull? The probability is that skull and jaw were once united. Both specimens differ in certain anatomical features from human bones and in others resemble the skull and lower jaw of a chimpanzee. For this reason the skull was not referred to *Homo sapiens* but called by a new name *Eoanthropus*, that is, the Dawn Man. While we should not be justified in speaking of *Eoanthropus* as the long-sought missing link, it may be described as the skull of a primitive type of human being and one of the oldest records of prehistoric man so far discovered. When did this caricature of *Homo sapiens* live? No precise answer can be given; it may have been before the beginning of the period described in this chapter, perhaps a million years ago or perhaps separated from the present by an interval covering less than half that time.

The records of pre-Glacial man are confined to implements, to his tools and weapons alone, and the chief cause of the unending controversy is the difficulty of drawing a distinction between pieces of flint which were chipped and rudely fashioned by man and those chipped and battered by Nature. Flints discovered in deposits considerably older than the earliest boulder clays, especially in East Anglia, bear the impress of human handiwork as though fashioned as instruments by which to make other things and thus suggesting a marked advance upon the intelligence of the higher apes. It is practically certain that man was in this country before the oncoming of arctic conditions and there is indisputable evidence of his presence during the more genial interglacial phases which preceded the last extension of the glaciers. Caves in the Vale of Clwyd have yielded flint implements associated with bones of lion, bear, hyaena, reindeer, woolly rhinoceros, and man. Kent's Cavern on the outskirts of Torquay affords evidence of occupation as a shelter from the time of early man through successive stages of the Stone Age to the Bronze Age and the Roman occupation. Another important fact is the discovery of human remains in old river terraces 50 ft. above the present level of those now being formed, associated with bones of musk-ox and reindeer, animals that still exist on the ice-free barren land of Greenland. Old banks of gravel and sand with the finer mud or alluvium deposited by rivers occur in many parts of the British Isles: in some districts, e.g. in the Thames valley, there are such terraces at different heights above the present river indicating changes in level since the Ice Age. At localities in the valley of the river Lea arctic plants have been found which indicate a relatively cold climate, and the occurrence of musk-ox and reindeer in another terrace near Maidenhead is additional evidence of the same kind of climate. Without enumerating long lists of fossil animals which can be found in books of reference, the points to be emphasized are these: when Britain was in much the same state as that of Greenland and Spitsbergen at the present day, arctic animals, such as the musk-ox, arctic fox, reindeer, and mammoth, lived in relatively ice-free places; when the ice retreated a warmer climate prevailed. In the interval between two glacial phases man and a host of animals including hippopotamus, lions, *Bos primigenius* (the *Urus* of Caesar) and many others were able to wander farther afield in a temperate and more genial England.

Another aspect of the Ice Age is its effect upon vegetation. Before the invasion of Britain by northern ice-sheets the flora was very similar to that of the present day, and so was the climate. It used to be thought that when the covering of ice reached its maximum the whole of the vegetation, except perhaps lichens and a few mosses, must have been destroyed or driven across the low-lying land which in pre-Glacial days united the British Isles to the European mainland across the North Sea and the Straits of Dover. The modern view based in part upon increased knowledge of arctic lands in both the New and the Old World is that throughout even the most intense glacial phase some of the hardier flowering plants survived both in the southern part of England and on areas in the Pennine Hills and on other ranges which were not overridden by ice. Several flowering plants which it is customary to speak of as arctic or alpine-arctic flourish on Scottish, English, and Welsh mountains generally high above sea-level. Beyond the southern limit of the Arctic Circle the same plants have a wide range and reach sea-level. From thin beds of peat interspersed among glacial gravel and sand, leaves, fruits, and seeds of familiar arctic plants such as *Dryas*, the mountain avens, and dwarf arctic birches and willows have been collected; they are reminiscent of the time when British valleys were occupied by glaciers. A single species of *Dryas* (*Dryas octopetala*, from *drus* = oak), characterized by its shrubby habit, its 8-10 white petals, and foliage resembling miniature leaves of oak, is abundant on Scottish mountains up to nearly 3000 ft. and occurs also on English and Welsh hills. In Ireland, as in Greenland to-day, *Dryas* grows at sea-level. This member of the Rose family is represented throughout arctic Europe, Asia, and North America. The effect of the Ice Age upon plant life, though not so disastrous as was formerly believed, was undoubtedly considerable; trees that flourished in the much milder pre-Glacial conditions were killed or escaped by slow migration to other homes; only arctic plants were able to survive. Bearing in mind that in Greenland there are nearly 400 flowering plants and a few ferns in districts beyond the edge of the ice-sheet, it is important to remember that several flowering plants have been collected from mountain tops protruding from the ice-sea as far north as latitude 81° N.

The above references to animals and plants are much too incomplete and sketchy to furnish more than the bare outlines of a reconstruction of life during the alternating phases of the Glacial period; they may none the less serve my main purpose—to demonstrate how the historian of the past bases his conclusions upon a wide range of facts. Physical evidence based on inanimate material, such as boulder clay, moraines, and many other relics of ice-action, is confirmed and extended by the fragmentary remains of animals and plants which serve as indications both of arctic and temperate conditions in the British Isles in the course of the many thousand years embraced by the Ice Age.

The southern extension of arctic conditions well into the Temperate Zone was not confined to the British Isles nor to the Old World: a large part of continental Europe was similarly invaded and enormous ice-sheets spread over North America reaching as far south as latitude 38° N. Many hypotheses have been advanced in explanation of this glacial phase: changes in the distribution of land and sea and consequent interference with oceanic currents, a much greater elevation of the arctic regions with ranges of mountains far loftier than any in Europe to-day which were the gathering grounds of vast snowfields whence ice-sheets spread far and wide over a North Atlantean continent extending from North America across the North Atlantic Ocean to Europe and northern Africa. The case for the existence of this great continent is fully stated by Mr H. E. Forrest in his book *The Atlantean Continent*; it must be admitted that his views are not accepted by most geologists. Much has been written on possible astronomical explanations of the Glacial period with special reference to the precession of the equinoxes and fluctuations in the intensity of solar radiation. These are only some of many attempts to grapple with a problem that still awaits solution. As Dr W. B. Wright says in his authoritative book, *The Quaternary Ice Age*: 'It must be admitted that, among the theories that have been brought forward to account for the phenomena of the Ice Age, there is not a single one which meets the facts of the case in such a manner as to inspire confidence.'

This chapter has been written in the hope that readers who have been able to read the foregoing pages may be induced to observe natural phenomena with a new interest, to see in things that all can see some of the more recent finger-prints of Nature from which it is possible to decipher a few pages of Earth's Story-book. The story of the Ice Age cannot yet be written in full: there is still much to be done; more and more collecting of evidence and intensive study are needed, and in this work amateurs can play a helpful part. A gravel pit, an exposure of boulder clay, an embankment of river gravel above the present level of the valley may all yield treasures to observant searchers. Boulders of far-travelled rocks that have hitherto escaped detection, fossil shells and bones in gravel pits and terraces—all may be of value. In order to find out what particular deposit is likely to yield results the first step is either to consult a geologist or look at the Memoir on the district published by the Geological Survey.

CHAPTER VIII

From the Glacial period to Britain as it is

During the later stages of the Glacial period the ice-sheets of the Northern Hemisphere gradually diminished in size; mountain ranges and hills emerged from their winding sheets and only glaciers were left in the valleys. As the climate became less severe they too retreated from the lower ground leaving records of their former position in crescentic mounds and ridges of moraines and in deposits of sand, gravel, and clay spread as fans by water issuing from the glacier snouts. It was in the course of this gradual dwindling of the ice that many of the fresh-water lakes, to which reference has been made, were caused by the damming of valleys by barriers of ice or glacial debris. In the earlier stages of the Ice Age the climate gradually became colder, so after the maximum had been passed there were corresponding climatic changes, but in the reverse order.

We have now to take up the story so far as it can be followed through the closing stages of the Arctic Age through the years which link the most recent geological past with the present. The facts on which the chronicle is based are derived from several sources: sheets or terraces of gravel and sand strewn by swollen streams over the ground vacated by the retreating glaciers; beds of peat on moorland and in valleys built up by generations of swamp-loving plants: remains of forests embedded in peat intercalated among layers of mud and silt deposited by rivers or incursions of the sea. From these and other sources it has been possible partially to follow not only fluctuations in climate but also in the relative levels of sea and land. There is reason to believe that before the beginning of the

Ice Age the land stood at approximately the same height relative to the sea as it does now; when after the lapse of many thousand years Britain had passed from a temperate to an arctic climate the land weighed down by a heavy load of ice was depressed. At the end of the Glacial period the land rose above the encircling sea. Evidence of oscillations in level is supplied by raised beaches and by old river terraces. On many parts of the coast it is possible to see isolated patches or long strips of shingle in places 10–25 ft. higher than the present level of the sea; these are undoubtedly old beaches formed by wave-action when the land was lower. To cite one example, the road from Newlyn to Penzance is on a raised beach or shelf. River terraces, often two or three in parallel series, are situated one above the other, the highest and oldest lying farthest away from the present stream; each terrace marks the former level of the water when it meandered over a higher and broader plain. Terraces occur 70 ft. or more above the present level of the Thames, others are at a height of 50 ft. or less and mark a later stage in the downward movement. If in the course of a railway journey one looks at the grass-covered slopes rising above a river valley, the step-like arrangement of the embankments of sand and gravel is a conspicuous feature. Flint implements, bones of extinct animals, and shells collected from river terraces have thrown light on the life of successive periods. It is also possible to trace the transition from early prehistoric man to the later stages of human history, to follow successive steps in the progress of man's development to an age at which he has risen to the rank of a skilled craftsman able to make highly finished and efficient implements and practise the art of pottery. Another important step was the cultivation of crops and the domestication of animals. In the older gravels and in the lower layers of deposits in caves roughly chipped implements and a miscellaneous assortment of bones are the only records of human occupation;

from the overlying layers and from burial mounds more skil-
fully fashioned implements occur in association with pottery
which by its form and ornamentation can be referred to well-
defined phases of culture. The discovery of bronze—an alloy
of copper and tin—marked the beginning of a new age;
weapons and utensils, though not entirely replacing stone,
afford evidence of a higher stage in development; and at a still
later date the occurrence of iron objects reveals the discovery
of the art of smelting iron ore. There is no sharp line between
one stage and the next, nor is it possible to assign accurate dates
to the steps in human development applicable to different
countries. In some parts of the world man is still living under
conditions not far removed from those of the older Stone Age
man in Britain. The human population of the British Isles was
still living in the later Stone Age when the historic period with
its written records had already begun in Egypt, Crete, and the
Near East.

We pass upwards from the time when the British Isles,
northern and central Europe and an enormous area in North
America were transformed from a temperate to an arctic environment, and after the lapse of several hundred thousand
years, during which the climate waxed and waned in severity,
we come to the dawn of the present age, when the land was
more open to the immigration of man, animals, and plants.
A comparatively few stunted arctic birches and willows survived the maximum extension of ice-sheets and glaciers on the
flanks and summits of the Pennine range, the Cleveland Hills,
on rocky islands scattered over the sea of ice, and on the strips
of land south of the Thames-Bristol line; these were the pioneers
of a gradually increasing company of immigrants from the
Continent, to which England was then united across the swamps
of the North Sea. Forests colonized the land that the Glacial
period had reduced to a treeless tundra; at first only the hardier

trees and shrubs were able to endure the still inhospitable conditions. As the climate improved other plants obtained a footing and at length the whole of the British Isles reached its present state. We shall first follow the changing plant population and endeavour to obtain a general idea of the methods which enable us to follow the later stages in post-Glacial history. Reference was made in Chapter VII to the flowering plant *Dryas* familiar to naturalists as a·characteristic member of our mountain flora and a favourite with rock-gardeners. In Greenland, where it flourishes through a considerable vertical range, its leaves and stems are carried by streams into lakes where they lie embedded in sediment. About fifty years ago a Swedish geologist discovered fossil leaves and other fragments of *Dryas* in beds of sediment in Ireland, Denmark, in the Alps and some other countries south of the Arctic Circle which had been deposited on boulder clay and were overlain by peat. With *Dryas* were found in some places fossils of the lemming, an animal characteristic of the present tundras and steppes of Russia and Siberia and formerly of the barren, dry tundra which marked the retreating ice of the Glacial period. The important point is that the layers of sediment containing the fossils, which included also dwarf willows and other arctic plants, rest on boulder clay left by the retreating ice, a fact demonstrating the existence in Britain immediately after the Ice Age of representatives of a vegetation at or near sea-level precisely similar to that which now occurs in the ice-free regions of Greenland and other Arctic countries. *Dryas* and several other plants lived in Britain through the Ice Age; they formed patches of colour on the sparsely covered moorland and tundra that were the dominant features of this country after the retreat of the ice. Thus it is that the name *Dryas* Clay was given to a deposit not uncommon on the Continent but rare in Britain which marks the initial stage of the post-Glacial period. *Dryas* and its companions on our mountains are relics of the age of ice; when the climate was still arctic and the ice-sheets had been replaced by glaciers they were able to grow at sea-level, but as conditions ameliorated and other plants less arctic in character invaded the low ground the Glacial relics wandered to higher stations where, with the exception of localities in western Ireland, they now remain. In order to follow the rebuilding of the British flora it is necessary to examine the thick beds of peat that in many places cover boulder clay and moraines. It has long been known that peat contains at different levels stems of large forest trees such as oak with their roots still attached to an old surface-soil. Their occurrence shows that swampy fenland was transformed into ground fit for trees: the fact that the buried trees are overlain by peat is evidence of a return to swampy conditions. The trees indicate an interlude—a change in physical conditions between two periods of peat formation. About forty years ago a great advance was made in the investigation of peat by which it has been possible to extract much valuable information bearing upon the history of post-Glacial vegetation. Professor Lagerheim of Stockholm conceived the idea of searching for pollen-grains in layers of peat at different levels, and at a later date the Director of the Swedish Geological Survey, Professor von Post, extended and improved the methods of his colleague. This brings us to what is called Pollen Analysis. Pollen is the name given to the dust-like material, often yellow in colour, which is scattered from the male organs (stamens) of flowering plants, trees, shrubs, and herbs, and from cone-bearing trees. Fortunately pollen-grains are efficiently protected by a highly resistant coat and those of many plants are almost indestructible. Some pollen is eagerly gathered by insects and they serve as unconscious agents of pollination, that is, the transference of the male to the female element in a flower; after pollination fertilization ensues. With very few exceptions the pollen of trees is not carried by insects but by wind, and it is mainly wind-distributed pollen that occurs in peat. Clouds of pollen dust are sometimes seen rising from beeches and other trees in spring when the inconspicuous flowers come to maturity. It is said that a single catkin of an alder produces four million pollen-grains and the large head of a maize plant fifty million. Glass slides smeared with a film of glycerine or some other sticky substance suspended on the roof of a building or exposed in an aeroplane at high altitudes collect many kinds of pollen together with the minute reproductive cells, or spores, of fungi. Pollen from forest trees is distributed by wind over the country-side, some falling on the ground, some into lakes and rivers. The pollen from forests that have long since disappeared can be recovered by a portable boring instrument from different levels of peat and associated material. Samples of the peat brought up by the borer are washed and treated with caustic potash or soda and then examined, after mounting in glycerine jelly, under the microscope: percentages of the various kinds are easily counted. Such samples do not supply complete records because the pollen of certain trees and shrubs, e.g. poplars, sycamore, ash, and yew, are not sufficiently resistant to be preserved; but the pollen of most trees and many herbaceous plants retains its structure almost indefinitely. The method has its limitations and the application of it is not entirely free from sources of error, none the less it is possible to reproduce a general picture of the vegetation, though it is not possible except in a few instances to distinguish one species of plant from another. On the other hand, genera can almost always be recognized owing to the fact that the pollen-grains have dis-tinctive surface markings; the grains from different trees have their own peculiar form and pattern. With the pollen occur many other plant fragments which furnish supplementary in-formation on the composition of the flora. As an illustration of the application of pollen analysis to the history and development of forests a brief account is given of results obtained by members of the Fenland Research Committee founded in 1932 at Cambridge with the object of compiling as complete a history as possible of the East Anglian Fens. The Fenland lies in a shallow depression bordering the western side of the Wash where the tall tower of Boston Church is a conspicuous landmark; it extends south past Spalding, to the east to King's Lynn, south to Wisbech and continues to within a few miles of Cambridge. The noble cathedral of St Etheldreda at Ely stands on a low eminence overlooking the southern Fenland. The area is occupied by peat associated with silt and clay resting in some places on patches of boulder clay and other deposits of glacial origin. Here we have a series of records of a long succession of events subsequent to the Ice Age, an epitome of post-Glacial history. The records are by no means complete: no deposits have been found containing remains of *Dryas* and other arctic plants that lived in the tundra after the retreat of the ice; the earliest phase is unrepresented in the Fens. It has, however, been possible to trace the changing character of the forest vegetation, oscillations of climate and of ground-level, occasional incursions of estuarine or sea water, and the occupation of sites by men of the Middle Stone Age followed by later immigrants of the Newer Stone Age, Bronze Age men and Romano-British folk up to the time of the present inhabitants. The Research Committee, including botanists expert in pollen analysis, geologists, archaeologists, geographers, and historians, has demonstrated the value of harmonious team-work in an undertaking involving many branches of natural knowledge.

There are two thick masses of peat, with occasional intercalations of water-borne sediment, separated from one another by a thick deposit of what is locally known as Buttery Clay, which marks an extensive flooding by a widespread incursion of brackish and salt water from the area that is now the Wash. From data obtained by the microscopical examination of samples of peat taken by the borer from different levels and from the thorough examination of beds exposed in excavations, many contributions have been made to our knowledge of the development of the vegetation as one phase in the physical conditions gave place to another. Skeletons and bones of animals, and remains of minute aquatic creatures enable us to visualize the changing scenes; implements of flint and pottery of many kinds provide clues from which to follow the development of human colonization. It is interesting to note that the drainage scheme inaugurated by Dutchmen brought over in the seventeenth century had been preceded by Roman efforts directed to the same end.

The following outline picture of the post-Glacial Fenland is based upon a concluding section, entitled *The Changing Face of the Fenland*, of a set of important papers by Dr Godwin of Cambridge. As already mentioned the story begins at a stage separated from the final disappearance of the ice by a few thousand years, at about 7500 B.C., when the Fenland as such did not exist; the floor of the depression was then made of clay upraised from the floor of a Jurassic sea more than a hundred million years ago, together with patches of boulder clay and glacial gravels. The aspect of East Anglia was similar to that of northern Europe and northern Canada at the present day; over most of the ground were woodlands of pine and birch with here and there in more favoured localities trees more at home in a milder climate. Wide river valleys much larger than the valleys of to-day wound their way through a sparsely forested region. Hunters of the Middle Stone Age who were able to travel over the bed of the North Sea left implements in the older layers of the post-Glacial deposits that have been reached in the deepest excavations (see Fig. 5). As the climate became more genial, elms and oaks with an undergrowth of hazel took the place of the pine and birch, and by degrees a new type of forest occupied most of the ground. Limes and alders secured a footing in a few places. At about 5500 B.C. there was a definite increase in wetness and a consequent spread of alders along river banks. The climate was also warmer; oaks, alders, and lime were the dominant trees. This, as Dr Godwin says, was the golden age of the post-Glacial Fenland: the tall, well-grown trunks of oaks buried in the peat are proof of the striking contrast between the forests of those days and the modern woodland of the Fens. River valleys filled with sedge fen, with clumps of alder and willow, were bordered by forest.. During the occupation by the New Stone Age man far-reaching changes occurred; the Fenland became water-logged either by an incursion of the sea or as the result of a wetter climate. The woodland was overgrown by reeds and sedge; stems of well-grown oaks embedded in the black peat bear witness to swampy conditions and the suffocation of the trees through lack of oxygen in the wet soil. In the New Stone Age the Fenland was a large tract of sedge fen invaded here and there by alder, willow and birch. The whole region was an inhospitable swamp. Gradually the climate changed to comparative dryness; oaks and in places pines and yews spread over the peaty ground, with hazel, dogwood, and buckthorn as undergrowth. Suddenly, and probably in consequence of the rupture of a coastal bar that had guarded the Fenland from the waters of the Wash, there ensued a widespread invasion by the sea which lasted long enough to deposit over the peat a thick mass of buttery clay. Eventually in the early part of the Bronze Age a covering of peat was laid over the clay and alders and willows flourished. The abundance of bronze tools and weapons shows that man wandered freely over the fens. At about 500 B.C. when the Iron Age superseded the Age of Bronze the climate grew colder and wetter; the lime was driven out and birch increased; the greater part of the district was uninhabitable. Another marine invasion occurred in the Roman period which left tracts of silt and salt-marsh. It was then that Roman farmers carried out extensive drainage work. So after an interval of a few centuries, during which the Fenland was left practically untenanted by man, we pass to historical records and read of the Fens as rich in wild-fowl and with fish abundant in the rivers and meres. Finally in the seventeenth century the Fenland was transformed by man's agency, and yet patches still remain reminiscent of the days when Nature had its own way. The long series of events, Nature's varying moods, changes in the landscape, man's progressive march through the centuries, carry us back to a remote past; on the other hand, in this unique English scene the past seems to be with us still. The flint-knappers of Brandon still practise the handicraft of Stone Age man; watch them at their skilful trade and look at the Cathedral in the Isle of Ely, the oldest part of which dates from 1083; we seem to be conscious of the companionship of generations of mediaeval craftsmen. The past and the present join hands and the enjoyment of a modern scene is enhanced by a knowledge of other days.

It is not only in low-lying Fenland that it has been possible to follow the march of events through the post-Glacial age: peat beds on hills far above sea-level have yielded much valuable information. At Blackstone Edge on the Halifax road in the Yorkshire Pennines diggings in the moorland peat disclosed well-preserved portions of a paved Roman road that had been laid on a peaty foundation probably about A.D. 120. At a lower level were found pieces of bronze and flint arrow-heads, relics of the Bronze Age; and at a slightly greater depth a horn-case of *Bos primigenius* with flint arrow-heads of the Newer Stone Age. From the microscopical examination of peat samples taken from different levels it was possible to form a general idea of the forest vegetation in a region that is now treeless. Finally below the base of the peat a thin layer of sand was reached containing tools left by men of the Older Stone Age. In peaty fragments scraped from the horn-case 146 pollen-grains were identified, including birch, oak, lime, and elm. The continuous growth of peat had already begun before the dawn of the Bronze Age converting forests into swampy ground occupied by a moorland vegetation such as we see to-day.

Many other instances might be given of the value of pollen analysis as a clue to the history of British vegetation. Reference has already been made to a former connexion between England and the Continent across the southern half of the North Sea: in addition to evidence furnished by old river terraces and other deposits of an alteration in the relative level of land and sea confirmatory facts have been obtained from material brought up in the nets of trawlers from the sea-floor. About 60 or 70 miles from the Yorkshire coast there is a submerged plateau not more than 60 ft. below sea-level rising rather abruptly from the deeper water on its northern edge: this is the Dogger Bank. From the Bank pieces of peat and wood have been dredged, also erratic blocks, bones of land animals including bear, wolf, hyaena, Irish elk, reindeer, woolly rhinoceros, together with a few objects of human workmanship. It is almost certain that when the lower part of the East Anglian Fenland was being formed England was not an island but was joined to the Con-tinent across the North Sea: the Dogger Bank was a patch of higher ground on a low-lying swampy plain. It was a relatively cold period, probably about 7500 B.C., when the estuary of a greater Rhine of which the Thames was one of several tributaries skirted the western edge of the Dogger Bank.

The forest that left records in the submerged North Sea plateau is one of a large number of submerged forests visible at low tide at many places on the British coast. Several such

forests have been discovered in excavations made in the course
of dock construction. One of the most impressive submerged

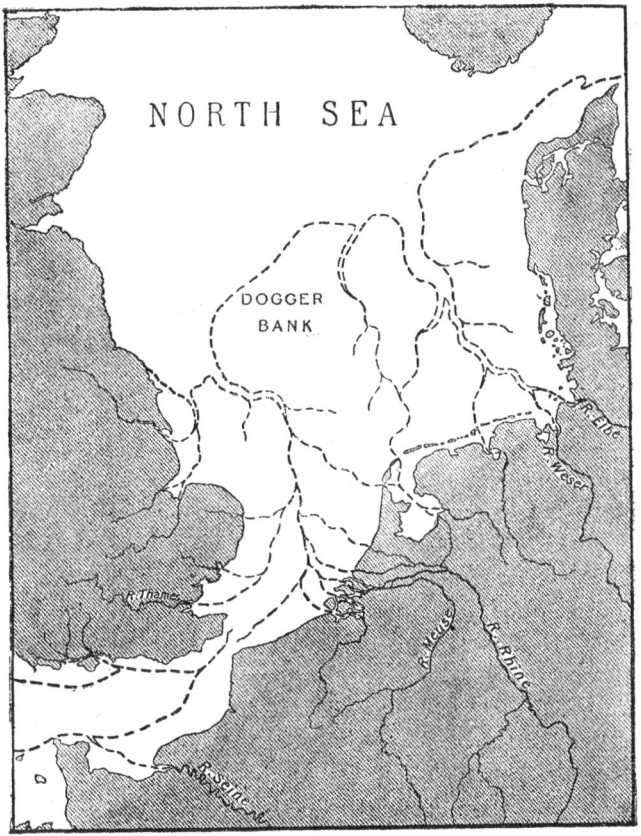

Fig. 5. North Sea coast-line at period of oldest submerged forest.
(From C. Reid's *Submerged Forests*. Camb. Univ. Press.)

forests covers a fairly large area off the Cheshire coast between
the Mersey and the Dee. At low tide at Leasowe it is possible
to see stumps of large oaks scattered over a peaty soil consisting

in part of a tangled mass of undergrowth stems of the Royal fern (*Osmunda regalis*) (Plate IV). There is no escape from the conclusion that when the trees were full of life the surface of the land must have been at least 70 to 90 ft. higher than it is now It has been suggested that submergence began at about 3000 B.C.

One of the many problems present to the minds of botanists is the origin of some of our commoner trees, whether, for example, they were introduced into Britain by the Romans or were among those which came naturally as immigrants from the Continent when post-Glacial climatic conditions were favourable to the spread of woodland. One example may be quoted: a few years ago the microscopical examination of charcoal with associated potsherds found in a prehistoric hearth at a locality near Cardiff showed it was beech wood and probably Iron Age in date, less than a century earlier than the Roman invasion. This is strong evidence in support of the view that the beech may be regarded as a native tree and not, as previously supposed, a Roman introduction.

As one scene succeeds another and our retrospect reveals contrasts in the landscape, the question inevitably recurs—how far is it possible to express in years the march of events? Accurate dating of stages in post-Glacial history is impossible: and yet approximate estimates are not altogether beyond our reach. It is certain that the Glacial period with its alternating periods of extreme arctic, moderate arctic, and temperate conditions lasted many hundred thousand years. Is it possible to say when the glaciers finally retreated? A few decades ago a Swedish geologist devised a method by which it has been possible to use as a time-scale layers of sediment left by streams issuing from the ends of melting glaciers in the valleys which they previously occupied. This ingenious method of counting the years, first employed in Sweden and later in other countries, has been generally accepted as the best available method of reaching approximate time-

PLATE IV

Submerged forest, Leasowe, Cheshire coast

measurements. In some localities which were formerly under ice there may be seen thick piles made up of thin layers of sediment lying one above the other in orderly sequence like fairly thick sheets of paper: the layers are in pairs, a layer of relatively coarse sediment and a layer of finer sediment. This regular alternation was interpreted as seasonal, each pair of layers representing a year; the coarser layer was deposited during each summer thaw when the force and volume of the glacier water was greatest; the finer layer was made of lighter particles transported by the feebler winter stream from the melting ice. The thin sheets of sediment, known as *varves*, can be traced for a considerable distance along the path of a dwindling glacier; they may be compared with a pack of cards lying at an angle on a table with the higher end of each card facing the upper part of the valley. At any one locality the pairs are counted and their thicknesses measured, then in a second pile farther along the valley it is possible to establish continuity with the first series by a careful comparison of individual layers or pairs of layers. By this comparative examination of sections through piles of sediment left by ice-streams Baron de Geer, the Swedish geologist, estimated that the ice finally retreated from southern Sweden approximately 12,000–14,000 years ago. It is not possible to make an equally satisfactory estimate of the corresponding event in the British Isles, but we shall probably be not very wide of the mark if we say that in this country the Glacial period ended approximately 15,000 or 20,000 years ago. Some authors hold that it was nearly 10,000 years ago, say 7800 B.C., when the northern boundary of the North Sea ran from northern Jutland by way of the Dogger Bank to what is now the Yorkshire coast in the neighbourhood of Flamborough Head. On the south coast the English Channel probably ended in a narrow bay off Beachy Head and farther east England and France were one. At about 5000 B.C. a general rise in water-level flooded the southern part of the North Sea and the land connecting Kent with northern France. Dates of more recent periods are based on various kinds of evidence furnished by comparative examination of collections of pottery and other archaeological material, the succession of forests revealed by pollen analysis and early historical writings in countries far in advance of Britain in human civilization. The Bronze Age is thought by some authorities to have lasted from about 2000 B.C. to 500 B.C., followed by the Iron Age, the Roman occupation, the Anglo-Saxon Age and so through the dark ages to the Domesday record of A.D. 1086. The Bronze Age brings us well within the historical period where Geology and Archaeology join hands; the archaeologist bridges the gap between the records of Nature and the records left by man.

CHAPTER IX

Changing Climates and Changing Life

In this chapter, reversing the order followed in Chapters VII and VIII, we shall take a general survey of geological history from the onset of the Ice Age to the lower limit represented by the surface of the Chalk. The first stage in our backward journey described in Chapter VIII, the records of the Ice Age, was considered before the more recent records in order to simplify the narrative: in this and the following chapters the reversed chronological order is more strictly followed. The rocks to be traversed are most of them sediments laid down in shallow water, some close enough to the land to contain remains of plants which furnish information about the nature of the vegetation on the coast or on river banks near the edge of the sea. These rocks are for the most part loose sands, clays, and occasional beds of peat; they are not the kind of material usually spoken of in common parlance as rock. The data upon which this volume of earth-history is compiled have been obtained from a series of old sedimentary beds of many kinds included by geologists in the Tertiary era; it comprises a period of time at least 60 million years in length. The whole has been subdivided for convenience of correlation and reference into subdivisions or stages with distinctive names, but for present purposes it is unnecessary to mention many of them. An outstanding feature of the Tertiary era is that the great majority of the fossils discovered in the considerable thickness of the various sedimentary beds are shells of animals that lived in the sea: the shells from the highest and therefore youngest beds are most of them specifically identical with existing animals; as we descend to lower and older sedimentary beds nearly all the fossils differ too widely from any found in seas of the present age to be identified with them and are clearly the remains of extinct forms of life. This gradual change from almost complete agreement with the present to an equally well-marked contrast in the fossils from older Tertiary rocks was used by Sir Charles Lyell in 1833 as the basis of classification: he called the later rocks of the Tertiary era Pliocene, the next older rocks he placed in the Miocene subdivision, and the oldest he called Eocene. These names from the Greek mean more recent (Pliocene), middle or medium recent (Miocene) and the dawn of recent life (Eocene). Fossils of Pliocene age include in some instances 95 per cent of species that are still living; the percentage of living or recent species in the oldest Pliocene beds is on the average 50. In the Eocene beds there are only 3½ per cent of living species. Subsequently as knowledge increased it was found desirable to intercalate a fourth subdivision designated Oligocene (few recent species) between the Miocene and Eocene.

PLIOCENE

What were the conditions in this country before the valleys in hilly districts were occupied by glaciers and, at a later date, when the greater part of the British Isles was covered by sheets of ice? It would be natural to expect some evidence of the approaching arctic climate in deposits immediately underlying the boulder clay and other kinds of material reminiscent of ice; and that is what we find. On the Norfolk coast near Cromer and elsewhere the cliffs are made of boulder clay containing enormous boulders of Chalk; below the boulder clay near the present sea-level a thin bed of clay containing fossil plants was discovered many years ago, and among the fragmentary samples of contemporary vegetation were the dwarf willow and the dwarf birch, stunted shrubs that now occur on the higher slopes of our mountains and throughout Arctic lands. Their occurrence indicates not necessarily an arctic climate but a lower temperature than that in East Anglia to-day. Below this so-called arctic bed there are sands and clays containing fresh-water shells, and below them more sediment with broken pieces of stems of forest trees. The abundance of tree stumps and logs was accepted as evidence that the bed exposed near the base of the cliff and on the beach at low tide was an actual surface-soil: it was therefore called the Forest Bed, a name that is still in use although it is now known to be a misleading description. None of the trees were found rooted in the ground where they grew; the wood had been carried by water as drifted debris from forests some distance away. From an examination of the wood and a large number of seeds, fruits, and other fossils it was possible to obtain a general idea of the vegetation at a stage immediately preceding in time the Arctic Plant Bed. With a few exceptions all the plants from the Forest Bed are members of the present British flora and clearly indicate temperate conditions very much the same as the climate of modern East Anglia. Bones of some thirty land animals were discovered in the Forest Bed, including wolf, hyaena, cave bear, hippopotamus, rhinoceros, elephant, the sabre-toothed tiger—a particularly fearsome beast—bison, musk-ox and others. Some of them are extinct, others are characteristic of countries warmer than Britain, but the musk-ox is now confined to Greenland and arctic North America. This company of animals is in marked contrast to the plants from the same bed practically all of which are still living and, with few exceptions, are represented in our present flora. England was at that time connected with the Continent across part of the North Sea then occupied by a low-lying plain with occasional lakes, a district comparable with the Norfolk Broads, over which the larger animals were able to reach Britain. The Norfolk sedimentary beds and their fossils throw light on the state of this country at a stage many years before the Glacial period; they tell us that the climate was already changing from comparatively warm to colder conditions, but the records are too few and incomplete to furnish more than a very partial picture. It is, however, interesting to note that the plants preserved in the Forest Bed indicate a climate practically the same as the present, whereas the larger animals were of a type very different from the present fauna of England.

Starting from the stage represented by the Forest Bed we pass to still older deposits included in the Pliocene period which gradually introduce us to a world more and more remote from the present not only in time but in the animal and plant population. The gravel and sand on the Norfolk coast containing bones of many kinds of land animals are believed to be part of the delta of a great river which flowed over the bed of the southern half of the North Sea: other examples of the same series of delta deposits occur in Holland and along the Dutch-Prussian border, but these samples do not belong to the same geological horizon, they differ from one another in the fossils they contain and thus make it possible to follow gradual changes in the vegetation on the banks of the river which, in the course of a few thousand years, was building up a delta stretching along the margin of the much reduced North Sea. The following figures illustrate the more noteworthy fluctuations in the character of the flora of western Europe up to the stage represented by the Forest Bed and the Arctic Plant Bed of Norfolk. Of 135 plants from the Cromer Forest Bed only 5 per cent belong to extinct species or species no longer living in Britain; the flora was practically the same as it is now. The examination of 100 plants from a slightly lower level in the old river delta in Holland showed that extinct and exotic species reached 40 per cent; out of 133 plants from a still lower level in the delta deposits in Holland extinct and exotic species reached a percentage of 88. It is clear therefore that during the time represented by the thick delta sediments the flora had become almost entirely transformed. Our knowledge of the floras is based upon seeds and fruits collected from the sedimentary beds. Few botanists are able to recognize more than a few seeds apart from the parent plants; in order to identify isolated seeds and fruits found in river-borne sand and clay it is necessary to examine a large number of specimens not only of British plants but plants now living in other parts of the world, including the Far East. This entails making a collection of seeds from herbarium specimens and other sources in order to obtain material with which to compare the fossils. It is only in recent years that the value of seeds and fruits as aids to the determination of plants has been fully appreciated.

Deposits belonging to the next older stage of the Pliocene period, below those already mentioned at Cromer and other places on the Norfolk coast, are best represented in East Anglia east of the Chalk, from Suffolk to Essex along the edge of the North Sea and inland to Sudbury and Ipswich and to Walton-on-the-Naze in Essex. In this East Anglian district there are thick masses of shelly sands, clay and pebble beds most of which were upraised from the bed of a shallow sea. The cliffs at Lowestoft, Southwold, Dunwich and other localities are made of these loose yellow and orange sands with layers of clay, locally known as Crag, all very similar to the material found in modern river deltas jutting into the sea where winds and tide pile up banks of shells mixed with sediment transported by rivers. The upper and younger series of these beds, known as the Red Crag, occur from Walton to Aldeburgh, at Woodbridge, Saxmundham and elsewhere; these deposits are rich in marine shells and were almost certainly formed in a shallow and turbulent sea. An interesting fact is that some of the shells of the Red Crag are closely allied to arctic species and were no doubt carried by currents from latitudes north of the English localities where the beds occur; many of the fossils are, however, species now living in temperate seas. The shelly sands next in age to those known as Red Crag and derived from an older sea are known as Coralline Crag, not because they contain fossil corals, but remains of marine animals known as Polyzoa (or Bryozoa), both recent and extinct species, some of which bear a superficial resemblance to small corals. The important point is that the marine fossils as a whole obtained from the Coralline Crag of Aldeburgh and other localities in Suffolk are comparable with molluscs and other animals now characteristic of warmer seas. As we descend from the younger to the older Pliocene sands the resemblance of the fossils to animals living in British or temperate seas becomes less and less. The old estuary of the greater Rhine, on the floor of which were deposited beds of sandy and other sediment containing many marine shells, overspread the land that is now East Anglia where the bays and inlets on the coast were silted up with shell-banks. The sea in which the upper and younger series of shell-sands, the Red Crag, was piled up near the coast-line was temperate while the older Coralline Crag sea was definitely warmer. No more bones of land animals like those of the Forest Bed occur in the underlying Crag deposits.

Evidence has been found in several districts showing that during the long period covered by the East Anglian shallow-water sediments the relative level of land and sea changed from time to time. The northern part of Britain was gradually sinking and the North Sea extended its boundaries: France and England were disunited. Patches of sand and clay contemporary with some of the East Anglian Crags occur in Cornwall at St Erth near Penzance; they are upraised samples of an old sea-floor and show that the land in south-west England formerly stood at a much lower level, about 400 ft. below the present English

Channel. There is evidence of this in platforms on the coast cut out of much more ancient rocks marking the position of former sea-beaches. The neck of land now connecting St Ives with Mounts Bay was a strait, and Cornwall with part of Devonshire was an archipelago of islands similar to the Scilly Isles at the present day.

MIOCENE

We have now reached a stage in the history of Britain where Nature's documents are lacking and there is a gap in the story, though, as we shall see later, not a complete gap. The events recorded in the series of beds from the beginning of the Glacial period down to the sands and other deposits known as Coralline Crag are comprised within the Pliocene period. Descending to a lower level in the geological table we come to rocks grouped together as Miocene, a period when a larger proportion of the fossils belong to species that are now no longer living. There are no representatives, except perhaps some pockets of sedimentary material in the Chalk of the North Downs in Kent, of this period in the British Isles and, in order to fill the gap, it would be necessary to consider the facts furnished by Miocene rocks and fossils on the Continent. All that need be said here is that an examination of fossil animals and plants from continental rocks shows a gradual amelioration in climatic conditions as compared with those in the Pliocene period. What does the gap in Britain signify? It suggests that conditions over the British Isles were unfavourable to the preservation of Miocene records because the whole country was above sea-level; there were no submerged areas where sedimentary rocks could be formed. Such gaps are not uncommon, mainly because most of the information used by geological historians is obtained from fossils preserved in old sediments laid down not on land but in fresh, brackish, or salt water. The absence of British beds of Miocene age does not therefore mean a complete gap in our history; from the gap, we draw the conclusion that while parts of the European continent were under water the British Isles were land. It is also possible partially to fill the gap from another kind of evidence afforded by clearly marked disturbances shown in the rocks at several localities in southern England. We know that in the course of the Miocene period events of great importance and stupendous magnitude occurred over a very large area in the Northern Hemisphere and, as we shall see later, these events left their impress upon British rocks.

OLIGOCENE

Meanwhile we shall take up the story as written in the sedimentary rocks included in the still older Oligocene period. In order to do this it is necessary to go to other districts of England than those in East Anglia where most of the Pliocene beds occur. Rocks of the Oligocene period and many included in the still older Eocene period occur in England in two areas known respectively as the London basin and the Hampshire basin. The London basin extends from Ipswich in the east through Hertford, Beaconsfield, Windsor and Reading, passing south-west to Aldershot and slightly north to Epsom, Croydon, Woolwich, Southend and up the Essex coast to Harwich. The Hampshire basin lies to the south of Salisbury Plain and the South Downs, from Dorchester on the west through the New Forest to Chichester and Worthing; it is a roughly triangular area with its apex near Salisbury, bounded on the south by the older folded rocks in the Isle of Purbeck and the Isle of Wight, and on the north by the arched ridges of Chalk and the Weald district. There is an outlying patch of Oligocene beds at Bovey Tracey near Newton Abbot in Devonshire. Within these areas are several kinds of deposits, some fresh water in origin, some marine consisting of sands, clays, and limestone.

At Bovey Tracey and Heathfield there is a thick mass of clay, which supplies material used in the local pottery works, associated with sand and a hard peaty substance, or lignite, consisting in great part of large and small pieces of wood. These deposits fill a hollow several hundred feet deep in the Devonian and Carboniferous rocks near the edge of Dartmoor. It is easy to collect pieces of fossil wood preserved as lignite in an open quarry near the road close to the Heathfield pottery. The hollow was once a lake into which rivers carried fine white mud derived from the disintegration of the granitic cliffs of Dartmoor; the lake was about 10 miles long and 4 miles broad and reached a depth of 600 ft. Embedded in the clay made of the white mud are seeds, fruits, twigs and other fragmentary remains of the vegetation on the hills and by the watercourses of an older Dartmoor, together with broken pieces of drift-wood. Some of the fossils are from plants that grew in swamps and fresh-water pools: overhanging the lake were forest trees with occasional festoons of creepers; other kinds of trees grew in ravines higher up the river valley. What was the nature of the vegetation bordering the lake and in the gullies cut by the river in its course from the uplands? There were ferns represented by fragments too small to be referred with confidence to living genera, but among them some that were closely related to the Royal fern, *Osmunda regalis*. One of the most conspicuous cone-bearing trees of the ravines was a *Sequoia*, a genus now confined to California and represented at the present day by two species, the mammoth tree of the Sierra Nevada Mountains and the more moisture-loving redwood of the Pacific coast: the Bovey Tracey *Sequoia* was most nearly allied to the mammoth tree, *Sequoia gigantea*, a commonly cultivated tree often spoken of by horticulturists as *Wellingtonia*. Among other cone-bearing trees were a yew (*Taxus*) and a swamp cypress (*Taxodium*): the living swamp cypress is now native in eastern North America and is occasionally grown in our parks and gardens. Fruits and other remains of a palm indicate a climate warmer than we enjoy to-day: there was also a *Magnolia* and a vine. There were some water-plants such as a pondweed (*Potamogeton*) similar to species that are now common in English rivers, a species of a floating water-plant related to our water aloe or water-soldier (*Stratiotes*). A noteworthy feature of the flora as a whole is that it was made up of species no longer living; the nearest existing relatives occur in North America, in Malaya and other distant regions.

One of the younger series of sedimentary beds in the Hampshire basin is seen on the north-west coast of the Isle of Wight at Bembridge, 2 miles south-west of Cowes, exposed in the cliffs at low tide. These beds, deposited in the delta of an Oligocene river, contain many plant remains along with bones of turtles and crocodiles with fresh-water and marine shells. The plants, largely represented by fruits and seeds with some twigs and leaves, belong to genera that are still living or to extinct genera; they point to a warm temperate or even a sub-tropical climate. Comparison of the vegetation with the nearest living relatives shows that the plants on the whole bear a close resemblance to species now native in eastern Asia and North America, a much smaller number being more nearly allied to European species. The only fern found in the Bembridge beds is closely related to one now common in swampy estuaries in nearly all tropical regions south of the Equator. In the Oligocene forest were cone-bearing trees that have long been known in this country only in cultivation; one was a conifer very closely related to the incense cedar of western North America (*Libocedrus*) which occurs also in eastern Asia and New Zealand; there was also a true cypress (*Cupressus*) related to the cypresses of southern Europe, and a species of *Araucaria*, a conifer now confined to the Southern Hemisphere, in Australia, New Guinea, the Pacific islands and South America. None of these cone-bearing trees is now native in Britain; all of them have long since migrated to distant homes and their modern representatives have been reintroduced to this country by man. The incense cedar is often grown in British parks and gardens; it is a tree with small scale-like leaves, narrow and tapering and similar to the true cypress and the Arbor vitae (*Thuja*): the best known Araucarias in cultivation are the South American monkey-puzzle and the Norfolk Island 'pine', *Araucaria excelsa*, one of the trees first described by botanists who sailed the southern seas with Captain Cook. Among the Bembridge flowering plants are several genera that are still found in European floras, e.g. a bulrush, a pondweed, oak and beech, a *Clematis*, a *Ranunculus*, and others: it is, however, important to note that these extinct Isle of Wight species are more closely related to plants now living in the Far East than to existing European species. One of the many flowering plants is a palm allied to a species now native in eastern North America, the West Indies, and South America. The significant facts furnished by a study of the plant remains in the fresh-water sediments of the Bembridge cliffs are: there is not a single plant specifically identical with any existing British species; most of the nearest living relatives are now native in China, Japan, and the Malay Peninsula; a few now occur in North America.

EOCENE

The next stage in our backward journey takes us to sedimentary beds, deposited many million years ago in a river estuary, and now exposed in the cliffs at Hordle (or Hordwell) between Becton and Milford-on-Sea on the Hampshire coast. From these beds many fossil plants and a few shells have been collected; we shall confine ourselves to the plants. The sediments are almost entirely fresh water in origin, made up of flotsam and jetsam carried by a large river probably flowing from the west into a wide estuary lying over part of the Hampshire coast in the Milford district and the western edge of the Isle of Wight; the occurrence of beds containing marine shells may be explained by occasional incursions of the sea in stormy weather over a bar which separated the fresh water from the open sea. Seeds, fruits, and pieces of wood transported by the river were preserved in the mud and sand of the estuary: some afford evidence of swampy conditions, e.g. the stemless palm *Nipa* represented by a species closely related to one that is now widely spread in tropical estuaries; a fern (*Acrostichum*) hardly distinguishable from a species commonly associated with *Nipa* in modern tropical swamps; also a species of *Stratiotes*, a genus of which there is only one living example, the water aloe of Europe and western Asia. It is interesting to note that this floating water-plant, now reduced to a solitary species, was formerly represented by several species and had a long past history; the sole survivor occurs in a comparatively few places in Britain, where presumably it was able to live through the Glacial period. Not a single flowering plant from the Hordle beds of Oligocene age is identical with an existing species: the flora as a whole is comparable with floras in south-eastern Asia, and 60 per cent of the plants are related to existing Malayan species; it is less closely allied to the present flora of eastern North America. The climate must have been sub-tropical.

Continuing our descent to a more remote age we come to a great mass of stiff bluish-grey clay, the so-called London Clay because 400–500 ft. of it underlie the London area; it reaches its maximum thickness of 500 ft. in Essex; in Wiltshire at its western boundary it is only a few feet in depth. The same clay is seen in the cliffs at Southend, Walton-on-the-Naze, and Harwich; the tunnel close to Ipswich station passes through it. It occurs also in Hampshire and the Isle of Wight; it lies beneath

London and through it run the tube railways; Windsor Great
Park and most of Richmond Park are on London Clay. On
the north coast of Kent in the Island of Sheppey the clay forms
cliffs 200 ft. high, and it was at Sheppey and Herne Bay that
most of the fossil plants were found on the foreshore. The
London Clay is a mass of sea mud uplifted from the floor of
a large estuary which once lay over the land that is now south-
eastern England; a large river flowing possibly from the south
carried in suspension an enormous amount of fine sediment
made from the scouring and disintegration of rocks, and on its
surface floated logs of driftwood, twigs, fruits, and seeds from
forests on the river bank; the flowing water bore along to the
dumping ground in the estuary bones of crocodiles, sea-snakes,
and the remains of many other animals that we now associate
with tropical countries. Almost all we know of the plants is
derived from fruits and seeds and not from leaves; fortunately
many of them have been petrified by the infiltration of the
mineral iron pyrites in consequence of which their structure is
often fairly well preserved. The fossil shells include specimens
of a *Nautilus* allied to the pearly *Nautilus* of the tropics described
by Oliver Wendell Holmes as

> Child of the wandering sea,
> Cast from her lap forlorn!
> From thy dead lips a clearer note is born
> Than ever Triton blew from wreathed horn!
> While on mine ear it rings,
> Through the deep caves of thought I hear a voice that sings.

The conditions in south-eastern England at the stage in the
Eocene period represented by the London Clay were no doubt
very much the same as can now be seen off the mouths of great
tropical rivers where the turbid water with floating drift-wood
can be traced as a dark band far into the clear ocean many miles
beyond the coast. If one looks at the clay of the Kent cliffs or

blocks of it excavated in the London area the chances are against the discovery of fossils, though many are scattered through the rock. Nearly all knowledge of the animals and plants is derived from fossils left on the beach; the clay is broken up by the waves and swept away, but the hard fruits, seeds, bones, and shells are left littered on the foreshore as records of a long-vanished age.

The investigation of the plant remains furnishes an admirable illustration of the possibilities of intensive and prolonged research. Mrs Reid and Miss Chandler, who have written a most valuable and scholarly account of the London Clay flora, collected more than 1000 specimens in the course of two short visits to Sheppey and Herne Bay. Before it was possible for them to identify the seeds and fruits which they collected and the material in the British Museum they had to spend much time in examining specimens taken from living plants and from herbarium sheets. Most of the London Clay species have been referred to existing families, but not to living genera; nearly all the genera are extinct and there is not a single existing species. Of 70 genera 47 per cent were found to be related to species now restricted to tropical lowlands. Five cone-bearing trees were identified: *Araucaria* that was also a member of the younger Bembridge flora, a species of *Cephalotaxus*, a genus now confined to the mountains of China and Japan, and others that no longer live in Europe. *Nipa*, the stemless palm also found in the Hordle beds, and other palms lived on the shores of the Eocene estuary. Among the flowering plants was a species of *Magnolia*, a genus now native in Asia and North America. It would be tedious to give a list of the numerous genera; it is the general character of the flora as a whole in which we are mainly interested. All the families recognized in the London Clay flora are now represented in the tropics; 11 per cent are confined to the tropics; 32 per cent are almost exclusively tropical; 46 per cent are equally tropical and extra-tropical in the present plant world;

only 11 per cent are mainly temperate in geographical range. The closest relationship of the flora is with 'the very heart of the East Asian tropics, namely, with the Malay islands, where 73 per cent of the genera are found at the present day'. The majority of the plants are most closely allied to genera now living in tropical rain-forests in the Indo-Malayan region.

In recent years much light has been thrown on the nature of the vegetation in southern England at different stages of the Tertiary era chiefly by a comparison of the fossil fruits and seeds with those of existing plants. Records of plant life contained in some sedimentary beds consist mainly of leaves often beautifully preserved in clay and other material. Good examples of leaves and other scraps have been collected from beds exposed in the cliffs of the Bournemouth district, at Alum Bay in the Isle of Wight, at Corfe in Somerset, and elsewhere. These fossils are rather older than those found at Hordle and slightly younger than the London Clay flora; they belong to plants no longer native in Britain and tell much the same story as the fossils already described, namely a tropical or sub-tropical climate in the southern part of what is now England. It would be tedious in this brief review of the main episodes recorded in Tertiary sedimentary beds to refer to all the many subdivisions recognized by geologists: in both the London and Hampshire basins there are fairly thick beds of sand and other water-borne sediment to which names have been given indicative of localities where they occur. One example may be mentioned: in several places in the London area, in Surrey, and in the Hampshire basin there are yellow and brown sands of shallow-water origin known as the Bagshot Sands; they can be seen overlying the London Clay on Hampstead Heath, Harrow Hill, Horsell Common, Poole Harbour and many other places.

Evidence of the effects of long-continued denudation of rocks of the Tertiary era is seen in many districts and there is no need to describe examples, but it is worth while to draw attention to an instance of the almost complete destruction of a sandstone which formerly occurred as a continuous set of beds. The world-famous monument of the Bronze Age at Stonehenge on Salisbury Plain is partially made of great blocks of sandstone known as Sarsens or greywethers containing only a few fossil roots and rootlets: the avenue and circle at Avebury in Wiltshire are entirely constructed of Sarsens, some of them said to weigh seventy tons. The name Sarsen may come from *sar*= troublesome, because the large blocks were a hindrance to clearing the land. The rock of which the Sarsens are detached blocks has never been found *in situ*; the original sandstone beds have been broken up by exposure to weathering agents into large pieces that are strewn over the Chalk in the neighbourhood of Marlborough and elsewhere. The precise age of these blocks used by the builders of Stonehenge and Avebury is probably Eocene; they are relics of a once continuous bed of sandy sediment, whether of fresh-water or marine origin is uncertain, that was no doubt widely spread over the Chalk of southern and south-eastern England. At Stonehenge there are also some smaller blocks of a different rock, crystalline and darker in colour and totally unlike any rocks known in the Chalk country. In 1923 chips of these foreign stones were microscopically examined and the structural characteristics were found to agree with those of an igneous rock in the Prescelly Mountains on the north coast of Pembrokeshire. This discovery, made by examination of the minute structure of the rock, solved a problem that had long been a puzzle, namely the provenance of the blue foreign stones at Stonehenge. The Bronze Age builders must have transported the heavy blocks a distance of more than 150 miles, probably most of the way on rafts by water, at a time when an arm of the sea reached far up the valley of the Avon close to Stonehenge.

As we travel farther into the past the vegetation of Britain becomes sub-tropical and even tropical, and its resemblance to that of the Far East gradually increases. It is of interest to note a marked increase in the proportion of woody plants in relation to herbaceous plants as we pass backwards from the present: the percentage of woody plants, that is, trees and shrubs, is now about 17; in the old flora of the Rhine delta to which reference has previously been made the percentage was 50. The flora of the London Clay is made up almost entirely of woody plants. In this connexion two facts may be noted: it is generally believed that trees are older products of evolution than herbaceous plants; also, tropical floras are richer in trees than in herbs. It must be borne in mind that the material preserved in sedimentary deposits gives but a partial picture of the vegetation in the mass; there must have been many plants in existence at any one period growing in places beyond the reach of rivers which were the agents by which samples of contemporary vegetation were preserved. None the less it is justifiable to assume that the main conclusions drawn from a comparison of fossil floras one with another and with recent floras furnish trustworthy information on the trend of evolution and climatic change.

Before giving an account of a British flora, rather older than that of the London Clay, which it is possible partially to reconstruct from fossils discovered in comparatively thin layers of sandstone and limestone in some of the Western Isles of Scotland, let us pass in review the geological background so far as it can be visualized at different stages of the Tertiary era. The geographical conditions in East Anglia in the Pliocene period were not very different from those at the present day; an arm of the North Sea lay over a large part of what is now Norfolk, Suffolk, and Essex. Proof of a greater and more widespread incursion of the sea is furnished by the Oligocene and Eocene sedimentary beds of the London and Hampshire basins. It is certain that these beds are merely remnants of a much more widely distributed series of fresh-water and marine sediments which happen to have been preserved in the protected basins where they now occur. In all probability southern England was covered by sea, sometimes an open continuous sea, reaching at least as far south as Paris; sometimes, as fluctuations in level occurred, the sea was replaced by low-lying land with estuaries and lagoons. When we descend to rocks below those at Bembridge and Hordle there is evidence of a greater spreading of the sea in south-eastern England and part of Belgium and France. The present separation of the Tertiary beds into disconnected areas, the Hampshire and London basins, is the result of movement of the earth's crust which caused folding, depressing the rocks in some places, elevating them in others; gradually the upraised portions were worn away by denudation and the basins left without their former connexions. If the layers of coloured sands and other sedimentary beds exposed in the cliffs overlooking Alum Bay be carefully examined it is easy to see that they do not lie horizontally, as they must have done when spread out over the floor of a sea or lake, but stand almost vertically like books on a shelf. These Alum Bay beds include London Clay and younger sediments, but all of them older than the plant beds of Bembridge. The Chalk cliffs, older than any of the rocks so far mentioned, of Freshwater Bay and other places in the Isle of Wight often afford very distinct proof of tilting and folding as shown by the inclination of the rows of dark flints which were originally arranged in horizontal layers. In many parts of England where Chalk is exposed similar signs of folding are clearly seen. When did this folding of the rocks occur? When beds belonging to a certain geological stage are folded and others, more recent in origin, are not folded the inference is that the tilting and disturbance occurred after the formation of the older rocks and before the deposition of the undisturbed higher rocks. As a result of the examination of rocks belonging to the Tertiary era, not only in England but in other parts of Europe and in Asia, it has been definitely established that the folding of the beds in the Isle of Wight and other places in southern England is merely the outer ripple of what has been called the Alpine storm, an expression indicating the gigantic scale of convulsions in the earth's crust over a very large region which reached its maximum during the Miocene period, a period which is practically unrepresented by sediments in Britain. The British Isles were then part of the vast continent which included the Arctic regions; Scotland, Greenland and Iceland were all one. To the south of this continent a broad ocean lay over southern and part of Central Europe, stretching far to the east across Persia and northern India; this transverse ocean is known to geologists as the Tethys Sea (see Fig. 6). The Alps of Switzerland, the Pyrenees, the Carpathians, and the Himalayan ranges had not yet been formed; the broad belt of the earth's surface where they now stand as a series of transverse ribs and curving festoons was submerged under the Tethys Sea. It requires no little effort of the imagination to follow the major events which ultimately led to the birth from the oceanic waters of the trans-European and trans-Asiatic mountain chains, and to accept a statement of them, not as a flight of fancy, but based on deduction from observed facts and their interpretation by thoroughly competent geologists—men who have made an intensive and prolonged study of the structure of the region involved in this amazing storm, which swept as a flood of irresistible power through thousands of miles of the rocky crust. Into the Tethys Sea rivers from the continents on its margin aided by oceanic currents had been transporting all kinds of sediment derived from the wearing away of the land, and in the deeper parts of the great trough, chalky ooze, with its miscellaneous collection of shells, including the large Foraminifer *Nummulites*, had been slowly accumulating. All this material, several miles in depth, lying on the slowly sinking floor of the Tethys lay over a foundation of much older rocks. At last the earth's crust, which for millions of years had been in a relatively quiescent and stable state as a reception area for the 'dust of continents to be', under the strain of the enormous load began to yield to forces that had long been under control; the floor of the trough began to rise. This upward movement began long before the Tertiary era but did not reach its culmination until the Miocene period. The position was this: two land masses, Europe on the north and northern Africa on the southern edge of the dividing sea, were relatively rigid and resistant portions of the earth's crust; between them the Tethys Sea partially filled with its beds of sediment was a much weaker and less resistant region. The pressure which was the initial cause of the great convulsion operating from the south rucked up the weaker Tethys belt into broad folds and troughs, squeezing them upwards and onwards towards the northern continent; folds, at first gentle and symmetrical, as they were pressed against the barrier in front of them were transformed into overfolds and recumbent folds, that is, the sides of the arches on the side whence the pressure came were forced forward and bent so that the folds became asymmetrical and the arched layers assumed the position of beds in folds that were almost horizontal instead of vertical. The rocks cracked and split; the harder and more brittle rocks, unable to respond to the pressure by regular folding, were split into strips and flakes which were thrust forward as wedge-shaped masses inextricably intermixed with the folded and overfolded layers of more pliant material. The older foundation stones below the Tethys sediments became molten or semi-molten under the influence of intense heat accompanying the crustal movement; in their plastic condition they penetrated along the cracks and fissures lying in their path and thus were

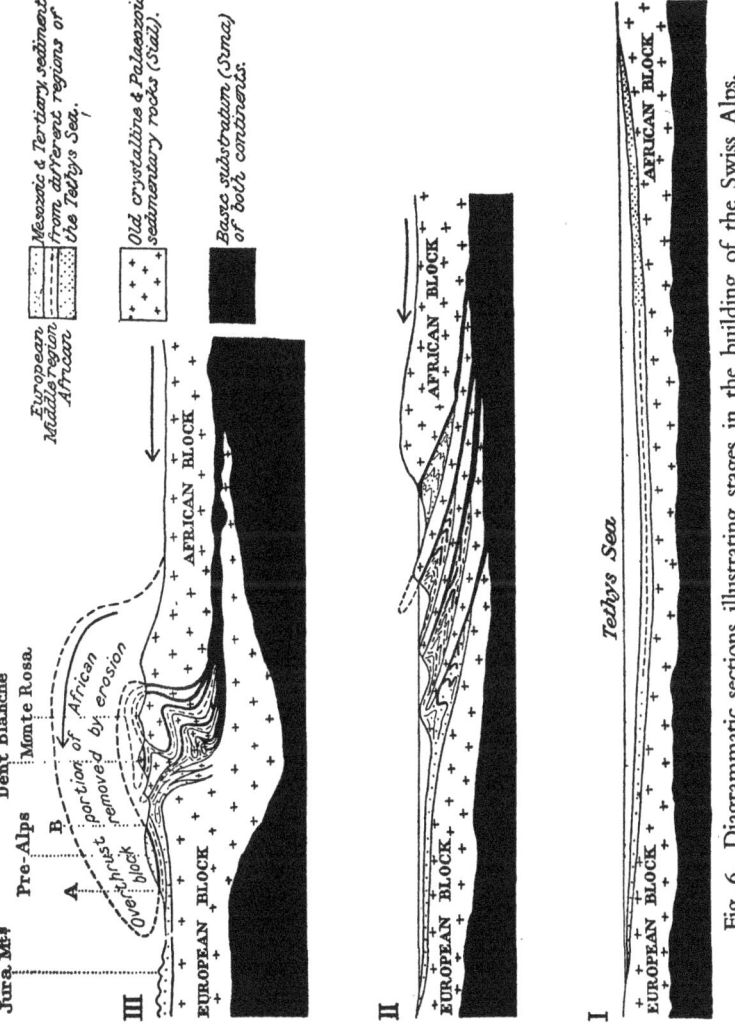

Fig. 6. Diagrammatic sections illustrating stages in the building of the Swiss Alps.

involved in the rising complex of rock sediments of many periods and intruded sheets and tongues of igneous rock. In the confusion caused by this upheaval and chaotic disorder older rocks were forced over younger rocks and came to rest in the reverse order of geological age. Large blocks of the African continent driven across the Tethys area by the continuing pressure from the south, were eventually piled up on the margin of the European continent; Africa made contact with Europe. This bodily transported block of foreign rock extends over several miles of western Switzerland on either side of the head of Lake Geneva, the district known as the Pre-Alps; from it have been chiselled the Matterhorn, the Weisshorn and other Alpine giants. Evidence of the truth of this almost unbelievable story is derived from several sources: the comparison of the rocks and their fossils of the Pre-Alps with those of the other Alpine districts; the occurrence of older beds lying above younger ones; the striking instance of violent folding seen on the face of many mountains; the recumbent folding brought to light during the boring of the Simplon tunnel. The pyramid of the Matterhorn in its upper part is built of crystalline rocks much older than the Jurassic beds at its base; these rocks are portions of once continuous sheets fashioned by Nature into the disunited mountains of the Swiss Alps. The origin of the Himalayan ranges was part of the same cycle of mountain-building which gave birth to the European Alps; the history of the building and upheaval of these two series of mountain chains begins with the piling up age after age of sediment on the floor of the Tethys Sea; but as the successive stages of their formation are followed many differences become apparent. The highest mountains in the world to-day are all relatively very recent: they have lost only a little of their pristine grandeur and eminence through the degrading and levelling processes of erosion. On the other hand the oldest hill ranges in the world,

PLATE V

North

South

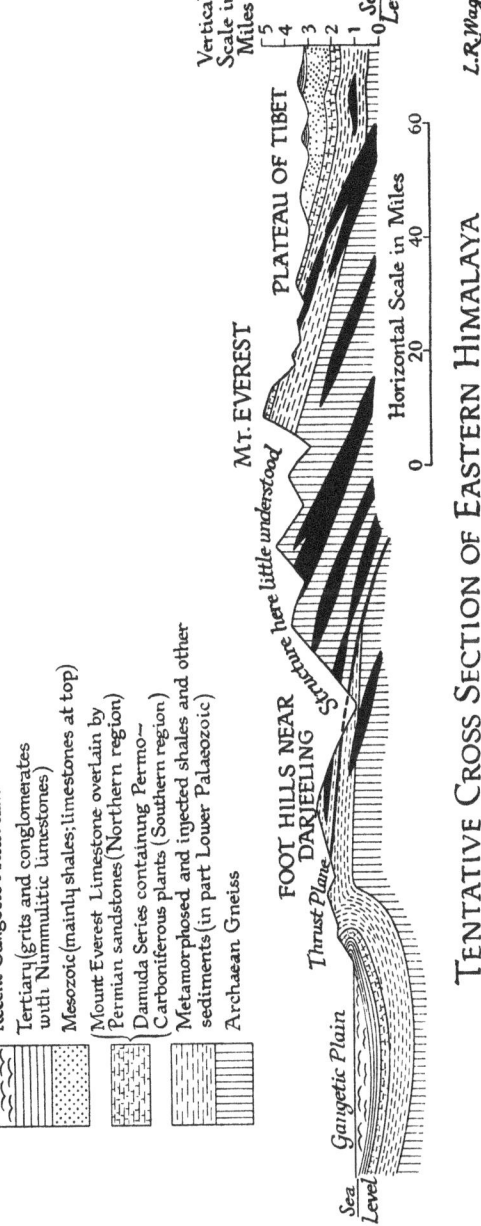

TIME SEQUENCE OF THE ROCKS

Recent Gangetic Alluvium

Tertiary (grits and conglomerates with Nummulitic limestones)

Mesozoic (mainly shales; limestones at top)

Mount Everest Limestone overlain by Permian sandstones (Northern region)

Damuda Series containing Permo–Carboniferous plants (Southern region)

Metamorphosed and injected shales and other sediments (in part Lower Palaeozoic)

Archaean Gneiss

Tertiary Granite Intrusions

Vertical Scale in Miles

5
4
3
2
1
0   Sea Level

PLATEAU OF TIBET

Mt. EVEREST

Structure here little understood

FOOT HILLS NEAR DARJEELING

Thrust Plane

Gangetic Plain

Sea Level

Horizontal Scale in Miles

0   20   40   60

TENTATIVE CROSS SECTION OF EASTERN HIMALAYA

L. R. Wager

such as those in the North-West Highlands of Scotland described in Chapter XVI and the equally ancient Aravalli range in India, are insignificant roots of mountain chains which may well have rivalled in height the Alpine peaks that pierce the clouds in the present age. These oldest hills are made entirely of pre-Cambrian rocks, including the oldest known parts of the earth's crust, and have now been re-exposed to the destructive action of Nature's sculpturing tools. The Himalayan ranges are made of a foundation of pre-Cambrian (Archaean) rocks, but above them and intermixed with them, as in the Swiss Alps, are great thicknesses of sedimentary rocks, often very highly metamorphosed by heat, ranging in age from the older Palaeozoic to Eocene. Beds of limestone containing the widely distributed *Nummulites* demonstrate the enormous area of the Tethys Sea over which that chalky ooze was formed; now it occurs several thousand feet above sea-level. The summit of Mount Everest (29,002 ft.) is made of limestone of Upper Palaeozoic age; the rest of the mountain consists of sediments of other ages together with crystalline rocks of the pre-Cambrian era (Plate V).

An adventurer into the Himalayan ramparts wrote: 'Looking at Himalayan peaks one seemed to be above everything in the world, and to have a glimpse almost of a god's view of things.' Knowing something of the construction of the ridges and soaring peaks, and of the circumstances of their birth, we think of them in their dignity and aloofness as earth's noblest memorials of the epic of creation.

Thus were born the mountain ranges that lie athwart the two continents: this was the Alpine storm. Thick masses of limestone that were once a soft white ooze on the bed of the Tethys Sea occur in the Swiss Alps 10,000 ft. above sea-level and in the Himalayas at a height of 20,000 ft.; this rock is largely made up of the small shells of a marine creature called *Nummulites* belonging to the class Foraminifera (Chapter III, p. 40). The name was suggested by the resemblance of the circular flat shells, sometimes about an inch in diameter, to coins (Lat. *nummulus* = a coin): *Nummulites* still lives in the Pacific Ocean. The age of the limestone is early Tertiary, rather older than the London Clay. Much of modern Cairo is built of nummulitic limestone. The Pyramids of Gizeh on the edge of the desert are built of great blocks of the same stone, and if these are examined with the naked eye or a pocket lens one sees flat circular discs of different sizes; many of them that have been worn or broken show that each Nummulite is made up of small chambers arranged in a regular spiral within the outer covering of the whole shell. Nearly 2000 years ago the geographer Strabo spoke of the Nummulites of the Pyramids as lentils. *Nummulites* occurs in some of the sedimentary beds of the Isle of Wight that are marine in origin. The folding and displacement of Tertiary rocks in southern England are correlated in time with the infinitely greater revolution that was changing the face of the earth in more southern latitudes.

We now return to western Scotland where the rocks tell a very different story. In order to present in true perspective the physical features in north-western Europe in the early days of the Tertiary era, it will be helpful to trace the events which began shortly after the land that is now the British Isles was uplifted from the sea, where for thousands of years the material that ultimately became Chalk had been slowly accumulating. The Chalk downs of England are only a part of the upraised white ooze from the floor of a clear and widespread sea; the same rock forms the lower portion of the cliffs on the Antrim coast and is hidden under newer rocks in western Scotland. At Garron Point on the Antrim coast early Tertiary lavas are seen to lie on Chalk and below both is the impervious Lias clay over which the great mass of lava and Chalk slipped to a depth of 150 ft. Following the great upheaval of the Chalk and the birth of a new land, subterranean forces that had long been held in check gained the upper hand: flows of semi-molten rock from deep reservoirs surged as a fiery deluge over the Chalk and over other and older rocks, converting thousands of square miles into barren lava-fields extending over an area not less than 2000 miles from south to north which reached far beyond the southern limit of the Arctic Circle. This unprecedented manifestation of volcanic energy, by no means confined to Europe and the Arctic regions, is one of the wonders of geology. It is convincing proof that after the lapse of many hundred million years the earth had not lost her youthful energy. There were three major phases of the age of fire: the first has left impressive records in some 3000 ft. and more of sheet after sheet of lava covering enormous areas, products of a series of outbursts either from deep fissures rent in the earth's crust under the compelling urge of subterranean forces or from scattered volcanoes of Hawaiian type. The columnar basalts of the Giant's Causeway, the columns of the 'Cathedral of the Sea' at Fingal's Cave on the island of Staffa, the basalts of Mull, Skye, Canna, Eigg and other Western Isles weathered into step-like terraces on the hill-sides, the flat-topped McLeod's Tables of Skye (1600 ft.), precisely similar basaltic platforms on the hills of Disko Island and the mainland of western and eastern Greenland—all these are parts of one stupendous whole, a plateau covering millions of square miles that was once a great northern continent. The second phase is represented by the more coarsely crystalline rocks of the Cuillin and Red Hills of Skye, and the granitic rocks of Goat Fell in the island of Arran: these rocks were not poured out as lava-streams, but were forced upward as a dome-like mass from a deep-seated subterranean source and, as their coarser texture proves, slowly cooled under the pressure of a thick superincumbent load: the comparatively large size of the crystals indicates gradual and not rapid solidification from a molten mass. These two phases of prolonged rock-building help us to appreciate the immensity of geological time. The second phase is an equally striking example of rock destruction as a measure of geological time. We see the jagged peaks of the Cuillin Hills rising to a height of 3000 ft. above sea-level which at no distant date, as earth-history is reckoned, were buried under a considerable thickness of younger rocks that have been utterly destroyed by the ceaseless operation of denudation. The world, to our limited vision, appears to be almost static; mountains we have thought of as symbols of eternity, seen through geological spectacles, take their place as episodes in a series of events which have moulded the changing features of the earth's face. The rocky covering of the world viewed by geologists, 'foreshortened in the tract of time', reveals itself as a dynamic, mobile crust responding from age to age to constructive and destructive forces which have operated since the earth's early youth.

As the long period of igneous activity reached its climax there followed a third phase, less spectacular though widespread in its effects: an enormous amount of molten rock was impelled upwards which did not flow over the ground as lava but forced its way into innumerable cracks and fissures in the earth's surface and there solidified as dykes, walls and sheets of basaltic rock (Chapter IV). Many examples of dykes running vertically through other rocks or inclined at various angles can be seen in the Inner Hebrides and in many other places; they vary considerably in breadth, from a foot or two to 100 ft., and by reason of their greater hardness often stand out conspicuously as projecting walls. One of the best places to see these dykes is on the beach near the southern end of the Isle of Arran where they extend outwards as dark bands in the red sandstone which they penetrated through roughly parallel fissures. Similar dykes of the same age occur also in northern England and afford evidence in their wide distribution of the great extent of country affected by the titanic forces employed by Nature's architect in the early stage of the Tertiary era.

So far the events chronicled in the igneous rocks of the Inner Hebrides, northern Ireland and other more distant regions have been spoken of as though there had been continuous outpourings of lava with occasional showers of ash and, in some districts, upwelling of gigantic masses of molten material which remained hidden below the surface until, in the course of time, the covering rocks were removed by erosion. There is, however, clear proof that the extrusion of lava was intermittent: intercalated among the lava-beds are layers of sedimentary material, hardened sand and mud, layers of coal, and beds of fine-grained limestone containing beautifully preserved leaves, some fruits and other plant fragments, also a few examples of insect wings and shells. The richest plant-containing layers occur near the base of the pile of basaltic lavas on Ardtun Head, the low 'headland of the waves' near the south-western corner of Mull, the island on which, from his home on Iona, Saint Columba must often have gazed. Trees, shrubs and other plants were able to colonize 'the frozen fires of Vulcan' during quiescent intervals: rivers flowing in channels cut in the lava-field carried, with sand and mud, scraps of vegetable debris, and it is from these water-borne sedimentary rocks, sealed and protected by later flows of lava, that it has been possible in some degree to reconstruct the vegetation which flourished for a time on the southern edge of the great northern continent some thousands of years earlier than the stage in geological history represented by the London Clay.

Almost all the plants obtained from the sedimentary beds among the lava-flows are trees: only one fern has been identified, a species of *Onoclea*, a genus now represented by the sensitive fern (*Onoclea sensibilis*), which is widely distributed in Canada and the United States; it occurs also in northern China and Japan. *Onoclea* is no longer wild in Britain but is often grown in our gardens. It is interesting to find that *Onoclea* formerly lived in north-western Europe and in Greenland, regions where it long ago failed to survive. Its present discontinuous distribution in North America and in the Far East is explained by the fossils in Mull. In all probability *Onoclea* originated on the northern Tertiary continent, perhaps north of the Arctic Circle, whence it spread radially into America, Europe, and the Far East; in the central or European region it became extinct, sharing the fate of many other plants that failed to survive the rigours of the Ice Age. Among cone-bearing trees were a *Sequoia* closely allied to the Californian redwood, and a *Cephalotaxus* nearly related to a species now living in Japan and China. Fossil twigs and cones found in the island of Skye agree closely with the existing *Cryptomeria japonica* of China and Japan; it is the tree of the famous Avenue of Nikko. One of the most interesting trees that left exceptionally well-preserved leaves in the Mull sediments is a species of *Ginkgo*, now represented by the maidenhair tree, so called because the leaves are similar in form and venation to the leaflets of maidenhair ferns. *Ginkgo* is one of the most impressive examples in the plant world of survivals from a remote past; the solitary living species, *Ginkgo biloba*, is familiar in cultivation, but it is very doubtful whether it exists anywhere as a wild forest tree. China was undoubtedly its last refuge and the oldest specimens are those in China and Japan. It is the last relic of a family that was once almost worldwide in distribution and played a prominent part in the vegetation in the Jurassic period. The leaves preserved in the Mull sediments, hardly distinguishable from those of the living tree, show that the maidenhair tree was a member of the British flora in the earlier part of the Tertiary era. The same species flourished at the same time in Greenland, Spitsbergen, and other

Arctic lands. One of the conspicuous trees in the Hebridean forests was a plane (*Platanus*) with large, handsome leaves and flowers almost identical with those of the existing occidental plane of North America. The only species now native in Europe is the oriental plane of Greece and Asia Minor: the familiar plane tree of our streets, the so-called London plane, is known only in cultivation and is probably a hybrid between the occidental and oriental plane. Among other trees in the northern forests were a hazel (*Corylus*) and a cornel (*Cornus*) similar to British species but more nearly akin to species now found in the Far East. There was also a vine and that, too, closely resembles a Japanese species. These are only a few of the trees and shrubs that grew in the forests of the northern continent from Mull in the south to high Arctic latitudes in the far north.

In the account of post-Glacial vegetation, reference was made to evidence furnished by the microscopical examination of pollen-grains preserved in peat: many different kinds of pollen have been found in layers of peaty material in the island of Mull, e.g. pollen of a *Magnolia*, a genus no longer represented in Europe.

Much more might be written on the fossil plants from Mull and other localities on the lava-fields of the northern continent that was long ago broken up into widely separated pieces; the important facts for our present purpose are as follows. The flora as a whole probably included no species still living in any part of the world, though it is difficult in some instances to say where the difference between the extinct and recent species lies. The flora is not, in a geological sense, very much older than that reconstructed from the fossils in the London Clay, and yet the two floras have practically nothing in common; the Mull flora is not tropical like the Indo-Malayan type of flora of the London Clay. Nowadays there are only minor differences between the vegetation of western Scotland and that of southern England.

Another point is that, while in both the northern and southern floras there is no single British species, there are several plants in the northern flora closely related to living British trees and shrubs. Moreover, the northern flora has its nearest living allies in the Far East, especially in the mountains of China. These well-marked distinctive features, even making allowance for the difference in age between the two floras, are noteworthy; they afford evidence of climatic contrasts. An interesting fact is that the flora of Mull agrees very closely with contemporary floras in Arctic regions, an additional proof of considerable change in climate between the early part of the Tertiary era and the present day.

We pass now to another question of no little interest to which it is not possible to give a very definite answer, namely, at what stage in the Tertiary era man made his appearance. The human race was undoubtedly in existence before the Ice Age, but how long before we do not know. Flint implements, now generally believed to be the work of man, have been found below the Red Crag of Suffolk: others were found in the Forest Bed above the Red Crag; the two deposits are separated by at least 50,000 years. The implements at the base of the Forest Bed may have been made by early man 400,000 years ago. The large size of these old implements, it is suggested, shows that they may have been fashioned by strong men with large hands. The probability is that animals, more human than ape-like, were witnesses of some at least of the events recorded in the delta deposits of the river which flowed from the Continent over the southern half of the North Sea, several hundred thousand years before the British Isles were as Greenland is to-day. Looking back through the mists obscuring the past, facts based on convincing evidence give place to fancies and assumptions; we see in imagination the gradual transformation of brute creatures into beings transitional between apes and man, creatures which, at a later stage in evolution, rose to a higher mental level and used their arms and hands for chipping stones into implements serviceable in the ordinary activities of life. The manner and date of this greatest of all steps in evolution are, as yet, only very partially revealed. It is at least certain that most of the drama of the rocks so far described was enacted millions of years before the birth of the human race.

It is difficult for us whose sense of time tends to be influenced, or even dominated, by the span of human history to form a true conception of geological time: a period that takes us back through 80 or 90 million years might be regarded almost as a measure of eternity, and yet the ages briefly considered in this chapter cover but a fraction of earth-history which began at least 2000 million years ago. One might expect that during this latter time there would be chronicled in the rocks only comparatively insignificant and orderly changes, events such as might be expected in the history of a planet that had been slowly cooling from a gaseous, molten state, changes consistent with enfeebled age. Since the discovery of radioactive minerals it has been necessary substantially to revise standards based on the conception of a gradually cooling earth. Loss of heat by radiation has been continuous, but there has been a no less continuous supply of heat from the disintegration of minerals within the rocks which rejuvenates the whole mass and, as it were, inoculates a senile earth with an elixir of perpetual youth. If this revised conception of the earth is correct, we should expect to find even in the more recent chapters of geological history evidence of persistent youthful energy. It is true that the happenings recorded in the rocks at the higher levels of the Tertiary era cannot be described as catastrophic or on a gigantic scale; they tell us of changes in the relative position of sea and land within comparatively narrow limits, of climatic fluctuations, of the existence of animals that no longer exist anywhere in the world, and of others that now live in countries remote from Britain. Descending to the lower levels described in the latter part of this chapter, we discover proofs of two very different, though probably not unconnected, far-reaching and tremendous revolutions in the structure of the earth's crust: the more recent was the conversion of the Tethys Sea into a chain of mountain ranges, including the highest peaks in the world. This Alpine storm, which transformed the face of the Northern Hemisphere long after the London Clay had been spread over the sea-floor, was geologically an event of yesterday. Going farther back to the older Tertiary rocks of western Scotland, we find another proof of the perpetual youth of a world over 2000 million years in age. The terrific convulsion that was the origin of the Alps and Himalayas did not spend itself in one great wave of mountain-building; there was an aftermath which found expression in the creation of a far-flung continent made almost entirely of igneous rocks, the products of long-continued volcanic activity and the uprising of molten material from subterranean sources. The continent was continuous from northern Ireland, the Western Isles, the Faeroes, Iceland, Greenland, Spitsbergen, and Franz Josef Land. Most of this continent lay beyond the British area, but the fossils collected from the sedimentary beds between sheets of lava in some of the Inner Hebrides enable us to visualize the astonishingly uniform vegetation represented in the forests and undergrowth which colonized the lava-field from the polar regions to the southern boundary of the northern continent of which Mull, Skye and other islands are dismembered fragments. The uniform character of this early Tertiary flora is clearly shown by a comparison of fossil plants long ago collected from rocks of the same age as those in the Western Isles, in Grinnell Land, Greenland, Spitsbergen and other Arctic lands. The close resemblance of many of the species still living in the Far East throws a flood of light on the wanderings of plants over the world's surface: many trees and shrubs on the mountains of China are, no doubt, the descendants of species that once flourished in the Arctic regions and on the western edge of Europe and may have been first evolved on the vanished northern continent whence they were driven by climatic change to seek new homes; those that migrated along a southern route into northern and central Europe were in all probability destroyed during the Glacial period, others that migrated to the south-east survived in the more genial valleys of China where they were not exposed to the devastating effects of the arctic cold.

CHAPTER X

The Flooding of the World

The chapter of history recording the physical and biological changes during the long series of years preceding the Tertiary era is called by geologists the Cretaceous period; it is so named because the Chalk (Lat. *creta*) rocks of England are an impressive memorial of an outstanding episode in the history of the British Isles in the latter part of the span of some 65 million years covered by the whole period. The Chalk is the most conspicuous and the most widely distributed rock of this period in Britain, but it is not the only kind of rock included in the Cretaceous age. In this chapter emphasis is laid on the major subdivisions of the period from which it is possible to follow the changing geographical conditions between the dawn of the Tertiary era and the closing stages of the Jurassic period. The Cretaceous series of rocks are grouped in descending order under the following names:

Chalk; Upper Greensand and Gault; Lower Greensand; Wealden.

The relative ages of rocks can usually be determined by the order of their arrangement as seen in a geological section, e.g. on the face of cliffs by the sea, in quarries and other excavations, the lowest beds being the oldest. When beds of different kinds succeed one another in regular order it does not necessarily follow that their formation as sediments was a continuous process, a continuous sequence in time with no interruption or pause. Rocks of Tertiary age are often seen to be directly above Chalk, and yet we know that in the British Isles there was a time-interval between the Cretaceous period and the Tertiary era. This knowledge is based on more than one kind of evidence: the uppermost part of the Chalk in England does not represent the latest stage in the Cretaceous period; in some localities in north-western Europe there are still higher and younger layers of Chalk below the oldest Tertiary beds: another piece of evidence is that the Chalk platform on which sands belonging to the earlier stage of the Tertiary era rest in English localities is uneven and shows signs of having been long exposed to denuding agents before the deposition of the younger beds. In western Greenland and some other regions there is a gradual transition from Cretaceous to Tertiary sedimentary rocks with no evidence of any break in continuity; in the British area, on the other hand, there is always a gap in time between the two sets of beds. What happened in Britain was this: the Chalk was upraised from the bed of a sea and became the surface of a new land; for a long time the Chalk land was exposed to the weather and much of it was worn away, also in places it came under the destructive influence of the sea. Eventually after the lapse of many years the deposits of the next age were deposited on the denuded surface. The land that is now England does not give us a complete story of the passage from the Chalk age to that which inaugurated the Tertiary era: there was an interval during which there was rock-destruction and not rock-construction.

In a previous chapter a short account was given of the extent of country where the Chalk forms the surface-rock or is hidden by superficial deposits such as boulder clay or gravel and sand. The present distribution cannot be taken as a measure of the area which it once covered: there is proof of its former occurrence over a considerable part of north-west Ireland where in the Antrim cliffs Chalk underlies Tertiary lava-flows; it occurs also in patches in western Scotland and there is reason to believe that Cretaceous rocks once covered a wide area in the heart of that country. There is something peculiarly English about Chalk: it is true that rock of the same kind occurs in some districts on the north European seaboard, but Chalk is essentially a characteristic feature of the English landscape:

> The long green roller of the Down
> An image of the deluge ebb.

Another writer on Downland speaks of the 'solemn slope of mighty limbs asleep'. Chalk makes a special appeal to our aesthetic sense; the thin green mantle over its smooth curves

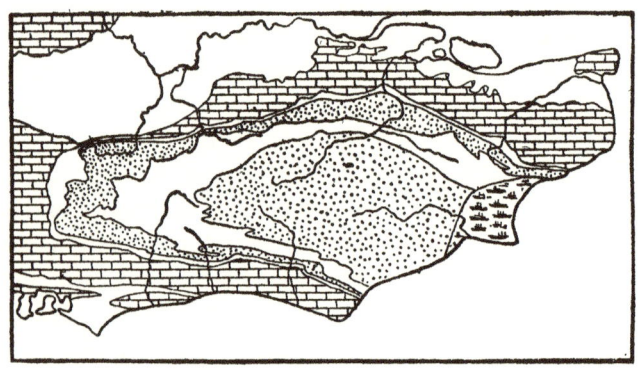

Fig. 7a. Map of the Weald.

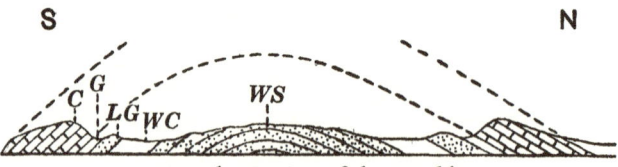

Fig. 7b. Section of the Weald.

WS, Weald Sandstone; WC, Weald Clay; LG, Lower Greensand; G, Gault and Upper Greensand; C, Chalk.

studded with flowers of the incomparable harebell and many other herbs, the bushes and trees that flourish on a soil rich in lime. The cliffs of Dover and the white walls on the Sussex coast have for centuries cheered the hearts of travellers returning home; narrow tracks over the Downs near Winchester and elsewhere, burial mounds silhouetted against the skyline, remind us of occupation by generations of pre-historic man. Even a slight knowledge of the nature and origin of Chalk adds substantially to our interest and pleasure by taking us back in imagination millions of years before the birth of the human race. In order to appreciate the full significance of the story revealed by the Chalk we must first ascertain what it is, of what it consists and how it was made. The thickness varies considerably, from about 1650 ft. downwards. In Norfolk it is 1400 ft. thick, under London 650 ft., at Beachy Head, near Eastbourne, 900 ft.; at Lublin, in Poland, a boring showed a thickness of 2800 ft. Not many years ago most geologists believed that the Chalk was formed on the floor of a deep sea: the chief reason for this was its supposed close resemblance in composition to a white chalky ooze that covers an enormous area of the ocean-bed at depths of 2000 fathoms and more. Recent research has modified this view: it is now believed that Chalk was built up through many thousand years as a deposit made of a white calcareous mud and the remains of shells belonging to a varied population of marine creatures in a comparatively shallow sea. It consists of about 90 per cent, or even more, of carbonate of lime, the substance precipitated as a white powder when air is blown through a tube into lime-water. There is a very slight admixture of particles of sand derived, no doubt, from sediment carried by currents or blown by wind from the land. The carbonate of lime is made up of from 5 to 10 per cent of minute shells of Foraminifera, the creatures whose remains make up most of the ooze, spoken of as Globigerine ooze from *Globigerina*, one of the living Foraminifera, on the ocean floor at the present day. Some of the Chalk consists of fragments of shells and complete shells of larger marine animals of many kinds, but most of it is finely divided material without recognizable structure, somewhat similar to a mud now being deposited through the action of bacteria in the sea off Florida and the Bahamas. Shells of sea-urchins, often well preserved, though rarely with the spines attached, are abundant fossils, also shells of bivalves of various sorts and rarely pieces of petrified wood carried by currents from the land. A striking feature is the abundance of flint usually in the form of irregular nodules a few inches in size but in some places in large blocks and more or less continuous layers. Flint consists of silica: occasionally on breaking open a nodule one finds remains of fossil sponges which had a flinty or siliceous framework instead of the horny skeleton of a bath sponge. We do not know exactly how flint was formed; some no doubt came from sponges, but probably more was deposited from sea water as jelly which became hardened into lumps of stone. Flints washed out after the removal by denudation or solution of the softer Chalk are the chief constituent of the gravel that is often seen on the upper surface of the Chalk. The possibilities of flint in decorative designs are illustrated in many of the fifteenth-century East Anglian churches; its use as a building material was realized by the Romans who constructed Richborough Castle near Sandwich. At a few localities small boulders of different kinds of rock have been discovered in Chalk and these may have been carried out to sea by floating seaweed or entangled in the roots of drifting tree stumps, or possibly even by floating ice. The conclusion is that Chalk consists mainly of finely divided calcareous material possibly produced by the action of bacteria which set up chemical reactions causing the precipitation of finely divided carbonate of lime; in part of shells and broken fragments of shells of many kinds of animals living in the sea far enough from the land to be out of reach of sand and mud that are deposited nearer the coast. The gradual piling up of all this calcareous material on the bed of a comparatively shallow sea until it reached a thickness of well over a thousand or even two thousand feet raises the question—how could that happen in a sea that was not deep? As the white ooze with the embedded shells increased in quantity the ocean-floor did not remain stationary but was slowly sinking, thus making possible a continued accumulation of the future Chalk without the sea being filled with its own deposit. It has been calculated that the ooze may have been formed at the rate of about 1 ft. in 10,000 years: be this as it may, the Chalk of Shakespeare's Cliff at Dover, Beachy Head and Flamborough Head represents several million years. The words put by Milton into the mouth of Eve when she talked to Adam—'With thee conversing I forget all time'—are not inappropriate to our contemplation of Chalk cliffs and the rolling hills of Downland.

From the present distribution of Chalk and other rocks included in the Cretaceous period we can reproduce in outline a picture of the geographical conditions over the area that is now the British Isles. There was land over part of Scotland and Wales bordering the Chalk sea, in Cornwall and Brittany; part of the Pennine Chain rose above sea-level and over the south-west of England was an archipelago of islands. An especially striking fact connected with the Chalk and other marine Cretaceous sedimentary rocks is the proof they afford that not only in Britain but in North America, western Greenland, North Africa, and as far away as Australia there was an almost world-wide flooding of the earth's surface. In North America a broad sea lay over the land from Alaska to the Gulf of Mexico covering the low-lying prairies; the Sahara desert was submerged and many other regions that are now land.

If we look at a Chalk cliff we can easily see whether the rock was uplifted from the sea vertically and retained its original horizontal position or whether it had at any time been involved in movement and folding of the crust; this information is given by the conspicuous rows of dark flint nodules which are characteristic of much though not of all the Chalk. The flints in the cliffs at Freshwater Bay and the Needles in the Isle of Wight, Ballard Cliff in the Isle of Purbeck and at many other places show that the rock has been tilted and now occupies an almost vertical position: this folding dates from the Miocene period when a very broad zone of the earth's crust was violently corrugated and fractured during the great revolution known as the Alpine Storm (see Chapter IX).

It is possible to some extent to appreciate the length of time represented by the thickness of the Chalk, which from its nature must have been built up very slowly on the bed of the Cretaceous sea, when we recall how Dr Rowe, whose work has already been mentioned, was able to follow steps in the evolution of sea-urchins by comparison of the species found at different levels. Evolution was a gradual process. When we confine our survey of animals and plants to the present world it is difficult to realize that each species marks a stage in a long developmental process: on the other hand, when we follow the procession of extinct organisms through hundreds of feet of Chalk and many other rocks a world that to us seems static and unchanging becomes dynamic and we see dimly the unfolding of life against the geological background.

Attention has been called to the almost complete lack of sand and other sedimentary material in the Chalk. A suggestion has been made, which illustrates how geologists endeavour to interpret evidence furnished not only by rocks in bulk but by microscopical investigation of small particles. If a piece of chalk is dissolved in acid the small amount of insoluble residue left may contain small grains of sand that are unusually well rounded, an indication that they have been exposed to the wearing and rubbing action of wind, possibly under desert or semi-desert conditions. Such sand-grains may have been blown by wind from sand-dunes and, though dunes are by no means confined to deserts, they are characteristic of dry wind-swept regions. If the inference is correct that rounded sand particles in Chalk came from the shore of a desert country, it affords another explanation of the lack of water-borne sediment, since there would be few or no rivers in a dry climate to act as transporting agents.

Chalk is not the only rock assigned by geologists to the Cretaceous period: immediately below the Chalk in many parts of England there is a sedimentary rock known as Upper Greensand, a sandy material containing in places harder and more compact layers of a flinty substance called Chert. The Upper Greensand lies on a thick mass of clay familiar to many people as Gault, a name of East Anglian origin; it is a muddy deposit upraised from the sea-floor as we know from the varied assortment of marine fossils. Gault forms the undercliff at Folkestone where it is pierced by the railway tunnel; it is seen in the cliff at Eastbourne and at many other places in South England and in the Cambridge district; it has been found in borings under London. With Upper Greensand it occurs in the upper part of the hill on the coast of Dorset known as Golden Cap, it forms the floor of many valleys as in the Vale of Pewsey, in Wiltshire, that is bounded by an escarpment of Chalk. The undercliff from Bonchurch to Blackgang in the Isle of Wight, made of a disorderly mass of fallen rock, is a good example of a landslip caused by the slippery surface of Gault underneath the Upper Greensand and Chalk. Similarly the landslip in the Axmouth district, near Lyme Regis, was caused by the sliding downwards of Chalk and Upper Greensand over the surface of the underlying Gault. The landslip occurred in 1839 when 40 acres and 8 million tons of rock slid towards the sea leaving a chasm over 300 ft. broad and about 200 ft. in depth. Two factors were concerned in these landslips: the slight inclination of the rocks towards the sea and the surface of the Gault made slippery by the accumulation of percolating water, that had passed through the permeable Chalk and Upper Greensand, on the impermeable clay. The Red Chalk familiar to visitors to Hunstanton, in Norfolk, as a bright red rock in the face of the cliff is a marine deposit older than the ordinary, white Chalk, and equivalent in age to the Gault. The important fact is that the Upper Greensand and the Gault are both marine sediments formed in a sea that was shallow and near enough to land to receive sand and mud transported by rivers and currents, in contrast to the clearer and deeper water in which, at a later date, the Chalk was formed.

Passing below the Gault, we come to the next older rocks that are called the Lower Greensand, another upraised pile of sandy sediment that was deposited on the sea-floor. In Cambridgeshire and Bedfordshire and some other districts the yellow and orange colour of many houses shows that they are built of Lower Greensand: the rock is in some places hard and in others of loose texture; a good example of the latter may be seen in the face of a cliff close to Sandy station, in Bedfordshire, which illustrates particularly well the irregular layering or bedding of the sand known as current-bedding (Chapter IV), a feature characteristic of sandy rocks deposited in shallow water with shifting currents.

So far the rocks we have traversed in this chapter are marine in origin: all contain samples of marine life and a relatively small number of fossils of animals and plants that lived on neighbouring lands. We now take up the Cretaceous chronicle recorded in the rocks of the Wealden district, that is, an area of hills, valleys, and plains over parts of Kent, Sussex, Surrey, and Hampshire (see Fig. 7a). The northern boundary extending from Farnham through Guildford and on to Folkestone is formed by the North Downs; on the south it is enclosed by the South Downs from Lewes to Beachy Head and along the Butser Hills near Petersfield in Hampshire. These two areas of Chalk, truncated on the coast at Deal and Dover on the north and at Eastbourne on the south, may be compared with the arms of an elongated horse-shoe, the open end of which faces the French coast, where the Wealden beds of England reappear beyond the present dividing sea bounded by a semicircle of Chalk that is the eastern counter-part of the Butser Hills. The Wealden area is geologically one of the most interesting and instructive districts in England; the rocks enable us to reproduce a fairly complete picture of the changing landscape in the early stages of the Cretaceous period and of the animal and plant population, also the effect of hard and soft rocks as factors that have been instrumental in giving to this small region its characteristic and varied scenic features. The oldest rocks of the Weald occur within the horseshoe; they are enclosed by elliptical bands of progressively younger rocks bounded by the youngest rock of all, the Chalk. Next below the Chalk is the Upper Greensand, and next below that is a broad band of relatively soft Gault sweeping as an expanse of low ground round the whole area and resting on a band of Lower Greensand concentric with it, on which are situated Dorking, Reigate, Sevenoaks and Maidstone. Passing farther from the periphery we come to a still broader zone of clay known as the Weald Clay: Haslemere lies on the junction of the Lower Greensand and Weald Clay. The central area of the Weald is made up of sandy rocks and clays with some limestone which form at the surface an irregular mosaic lacking the regular disposition of the encircling bands of younger rocks. On the higher ground at Tunbridge Wells weathered exposures of hard sandstone are a conspicuous feature; East Grinstead is situated on ground where Weald Clay forms the surface. A general view from a vantage point such as Leith Hill looking south shows an expanse of low-lying country occupied by the Weald Clay, including the valley of the Medway and Ditchling

Common. Escarpments of the harder and more resistant Lower Greensand are prominent features of a sandy district with heath and pine woods. Leith Hill (965 ft.) is made of Lower Greensand; St Martin's church stands on a Lower Greensand escarpment between Guildford and Godalming overlooking the flat Weald Clay. The hard sandstone below the Weald Clay joins higher ground culminating in Crowborough Beacon (800 ft.). The nature of the fossils from the sandstones and clays within the horseshoe and next below the Lower Greensand show that the beds are sediments deposited not on the floor of a sea, but in fresh water and not far from land. The Lower Greensand was deposited in a shallow sea in contrast to the fresh-water conditions clearly revealed by the fossil content of the beds that lie below it. The Wealden rocks are the upraised sediments of a large fresh-water lake covering south-eastern England and parts of Belgium, northern Germany and France; not many miles away was the western margin of an open sea separated from the lake by hilly ground composed of rocks that have long been buried below the surface, rocks that form a great ridge discovered by borings through the much younger beds lying above it. It was in this concealed ridge—a denuded range of Palaeozoic hills—that borings discovered coal in 1890 in the Dover area. Rivers flowing over the land on the edge of the Wealden lake built up deltas of sand and mud which gradually advanced farther and farther over the lake-floor; the lake sediments were subsequently uplifted to form the Weald Clay and the sandy rocks of Tunbridge Wells. As rock-building continued the floor of the lake gradually sank, making possible the accumulation of sediment to a thickness of over 2000 ft. After the formation of the sedimentary beds on the bed of the lake the area sank to a still lower level and the waters of the neighbouring sea swept over the district where the lake had been: in the early stages of the marine invasion the lake was comparatively shallow and in it were deposited the sands which ultimately became the Lower Greensand; later, as the depth increased thick masses of mud took the place of sand and these became the Gault; conditions then changed and the Upper Greensand was laid down; later still, as the depth of water increased the sea encroached over more of the land, the edge of which receded beyond the reach of detritus carried by rivers. Over the floor of the clear and deeper sea there slowly spread a white ooze containing shells of marine creatures of many kinds until after a long period the soft limy material was converted into Chalk.

By following the various beds of Chalk, Gault, Greensand and the older Wealden beds from the northern to the southern boundary of the Weald and noting the angle of inclination of the beds, the relation of the several kinds of rock one to another becomes apparent and a definite structural plan is revealed. A section across the district is shown in Fig. 7b: the whole area has the form of a broad truncated arch, the upper portion having been removed in the course of long-continued denudation. The Chalk of the North and South Downs is the basal portion of what was once a complete arch; similarly the escarpments of the Lower Greensand are remnants of the lower part of the series. Older rocks occupy the middle of the folded area. As the rocks of the Weald were being gradually folded they came under the influence of denuding agents, especially rain and frost: the country as we see it now with its varying surface-features, plains of less resistant and relatively soft clay, escarpments of Lower Greensand with steep faces sloping towards the centre of the horseshoe and the less steeply inclined surface sloping outwards. In the first stage of the upheaval the thick series of sedimentary beds of many kinds was arched upwards; in the subsequent stage of destructive erosion and denudation the different textures and hardness of the rocks determined the

pattern imposed by Nature's sculpturing tools upon the surface
of the Weald. In the central part of the area denudation has
laid bare a few patches of rock older than any members of the
Cretaceous formation. When were the rocks in the Wealden
district subjected to lateral pressure which caused them to be
elevated into the broad arch that is still clearly discernible in the
present structure of the foundation-stones? The folding was
accompanied by strains and stresses which fractured the rocks
at many places and caused local disturbances in the arrangement
of the beds, especially the sandstones and clays in the older
and central part of the district. In the Isle of Wight rocks of
Cretaceous age show equally striking evidence of folding and
fracturing. Geologists have satisfied themselves that the upheaval reached its maximum at a period a good deal later than
the London Clay. For the Miocene stage, which is almost
entirely unrepresented by sediments in the British Isles, has left
its impress in structural features imposed upon older rocks at a
time when mountain ranges on the European mainland and in
Asia were being lifted from the Tethys Sea. The British area,
though outside the region of most intense crustal disturbance,
was not entirely immune; the Wealden rocks, and the vertical
beds of Chalk in the Isle of Wight and elsewhere, are two among
many proofs that England was affected in a minor degree by
the far-reaching Alpine storm.

The fresh-water sediments of the Wealden lake are well
displayed on the east and south coasts of the Isle of Wight, in
the cliffs of Compton Bay, Sandown Bay, and Atherfield and
other places. On the foreshore at low tide, nearly half a mile
south-west of Brook, there can be seen a tree trunk 20 ft. long
festooned with seaweed and partially converted into stone by
the infiltration into the tissues of water charged with carbonate
of lime (Plate VI). The examination of thin sections showed that
the fossil tree is closely allied to the Scots pine; it was carried by a

PLATE VI

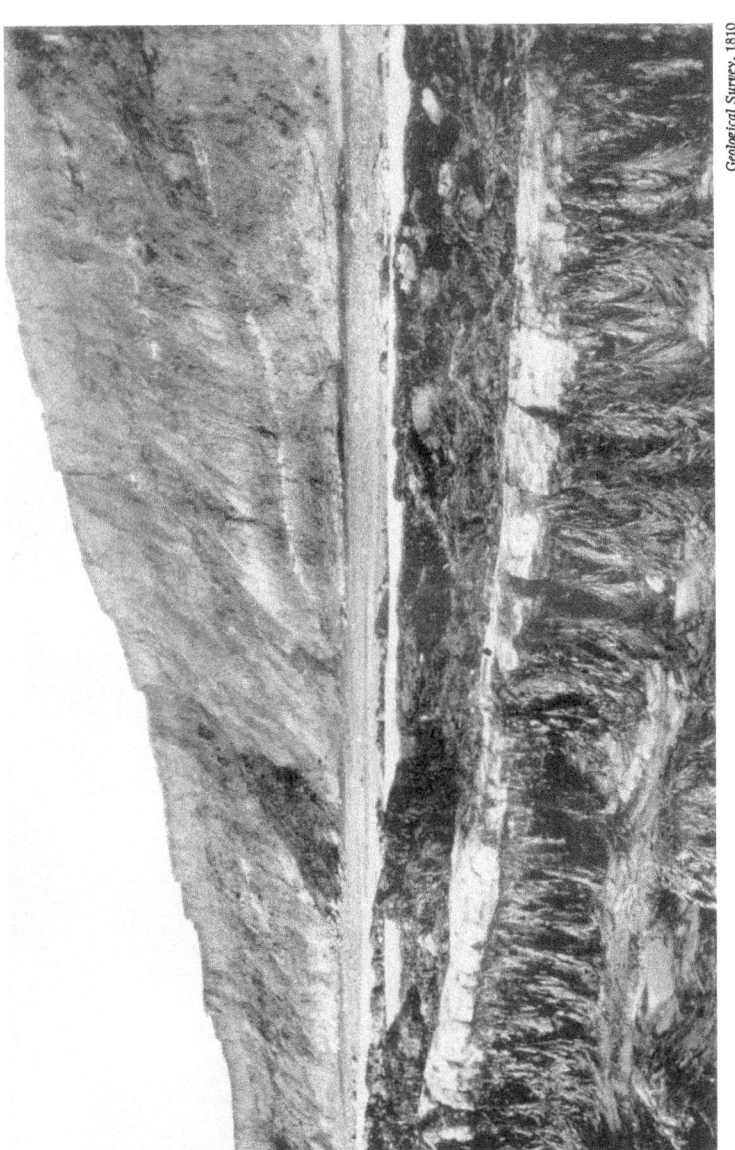

Geological Survey, 1810

Fossil tree at Brook, Isle of Wight

A petrified stem of a Wealden conifer, 20 ft. long

river from the woodland bordering the Wealden lake. The rocks which have been rapidly traversed in this chapter, from the Chalk to the sands and clays of the fresh-water lake, embrace a time-period extending over several million years. We shall now take a general survey of the records of life that have been discovered in the sediments from the Chalk sea and in those from the older fresh-water lake. Reference has already been made to fossils from the Chalk; they include shells of Foraminifera, a group that is well represented in the seas of the present day; much larger shells belonging to many kinds of molluscs similar to, but not identical with, living oysters and other bivalves, shells that were the external protective skeletons of marine animals belonging to another class that is now represented by the lamp shells (Brachiopods), characterized by two valves, one rather larger than the other, and differing from the shells of other bivalves (Lamellibranchs) in their more symmetrical structure. Sea-urchins were particularly common in the Chalk sea; there were also sea-lilies or crinoids closely related to species now living in both shallow and deep oceans; a few corals, shells of *Nautilus*. There were other spirally constructed shells, more elaborate in structure than those of the *Nautilus*, belonging to an extinct group known as Ammonites, so named because some of the shells resemble the closely wound horns of the Egyptian deity Ammon. Among other extinct creatures of the sea are Belemnites, common fossils in Cretaceous and Jurassic rocks, which are the hard internal skeletal parts of animals related to cuttle-fish; in old days the cylindrical pointed Belemnites, a few inches long, were known as Thor's thunderbolts. Sponges are among the more abundant Chalk fossils and their remains can often be found in flint nodules. Skeletons and grinding teeth of fishes are also common. The animals of the Chalk and those preserved in the Gault and Greensands, though none of them are identical with existing species, do not differ very widely from inhabitants of modern seas; they suggest comparatively warm water of no great depth.

In order to obtain a general idea of the land vegetation bordering the Chalk sea, we must go to other countries where plants have been found in shallower water, deposits approximately equivalent in age to the English Chalk containing vegetable debris drifted into deltas in river estuaries in North America, western Greenland and the continent of Europe. Practically all the fossil plants are twigs, leaves and other scraps from trees; we know hardly anything of the contemporary herbaceous plants. There were many ferns; among the more abundant are fronds of several species of a genus (*Gleichenia*) that is now almost entirely tropical and no longer exists in Europe. Cone-bearing trees were conspicuous in Cretaceous forests: trees closely related to the redwoods and mammoth trees (*Sequoia*); trees allied to the umbrella pine of Japan (*Sciadopitys*) which is cultivated in botanic gardens, ancestors of the maidenhair tree (*Ginkgo*) and several others. Broad-leaved trees include planes (*Platanus*), *Magnolia* and many other trees and shrubs which had foliage and flowers differing but little from those of living species. The more noteworthy features of the Cretaceous vegetation preserved in rocks corresponding in age to the Chalk, Gault and Greensand are as follows: they do not differ much from existing members of the plant kingdom, but they belong to trees which for the most part no longer live in northern Europe or anywhere in Europe; their nearest of kin are found in North America and Asia. In the latter part of the Cretaceous period the present dominant class in the plant world—the flowering plants—occupied a prominent position in the northern forests which flourished over a broad region reaching to within a few hundred miles from the North Pole. The climate must have been much warmer and much more uniform in the Northern Hemisphere than it is now: one of the most impressive examples of this contrast is furnished by the discovery in western Greenland and North America of remains of the bread-fruit tree (*Artocarpus*) that is now restricted to the tropics.

Passing to the older Cretaceous rocks represented by the sediments of the Wealden lake, we notice certain clearly defined differences. Ferns are still fairly abundant and comparable to species that are now tropical or sub-tropical. Cone-bearing trees include pines, araucaria, cypresses and others. Wealden rocks of northern France and Germany have yielded good specimens of extinct maidenhair trees. Some of the best fossil plants from which it has been possible to reconstruct the vegetation that grew on the shores of the Wealden lake were obtained in England from rocks at Ecclesbourne and Fairlight near Hastings. There were many kinds of plants known as Cycads, though many of them were only remotely related to living species: Cycads are now a small group almost exclusively tropical and nowhere abundant, some of them, e.g. in South Africa, are known as sago palms because their large fronds, several feet in length, are superficially very similar to those of palms; they are, however, not related to palms but are members of the large class of naked-seeded plants (Gymnosperms) which includes the cone-bearing trees. Most of the Wealden Cycads, as they are misleadingly called, belong to an extinct group which was formerly almost world-wide in distribution. An important point is that, so far, no flowering plants have been found in the older Cretaceous sediments in England. The vegetation of the latter part of the Cretaceous period, as we have seen, included many flowering plants such as planes, magnolias, the bread-fruit tree, and several others; there were some Cycads but not as many as in the older Wealden forests. This contrast raises an interesting question; flowering plants are now the largest class and the most widely distributed; they played the same dominant rôle in the Tertiary era and we can trace them farther back into the later stages of the Cretaceous period when, though less numerous and less varied than they subsequently became, they were still a conspicuous feature in the vegetation. On the other hand, when we pass to the early Cretaceous beds of Wealden age no flowering plants are found, and so far as we know in the still older Jurassic vegetation there were practically no flowering plants. The history of flowering plants is much too complicated and difficult to be discussed here; there are certain aspects, a brief reference to which may be of interest. When and where did flowering plants first make their appearance? We do not know the answer to either of these questions. The problem of the origin of the most successful class in the plant kingdom has long been present to the minds of students of evolution and many attempts have been made to solve it. We know that flowering plants of the kind we are familiar with to-day have been traced back through the Tertiary era and through the Cretaceous period. It is true that no examples of them have been found in English Wealden sediments, but a few are known from rocks in Greenland believed to be equivalent in age to our Wealden beds. Possibly, and indeed probably, ancestors of some existing trees were first evolved in Arctic lands and gradually spread farther and farther south. Flowering plants such as we know in the present floras of the world must have been preceded by ancestors long since extinct and probably bearing little resemblance in foliage and flowers to their living descendants. One fact is well established and that is the apparently rapid and sudden rise to dominance of flowering plants in the course of the Cretaceous period; the earlier history of the class is still shrouded in the mists of past ages. The search for origins is never-ending and one of the most fascinating occupations of those who seek to unravel Nature's secrets.

Attention was drawn in the preceding chapter to the abundance of large animals in Britain in the Norfolk Forest Bed, in deposits assigned to the Glacial and post-Glacial periods. The añimals such as the mammoth, rhinoceros, lions, bears and many others are members of the highest class, the mammals, so called because mothers suckle their young. Some of these Tertiary mammals still live in countries remote from Britain; some are extinct. The Tertiary era is often called the Age of Mammals as it was within that period and the following Quaternary period that mammals reached their climax in variety, size, and geographical range.. Skeletons and scattered bones discovered in Tertiary rocks older than the Norfolk Forest Bed have thrown an interesting light upon the evolutionary history of such animals as horses and elephants; the ancestors of our modern horses were no larger than a fox. Similarly, ancestral elephants are known to have been much smaller than the late Tertiary and living species (Chapter XVII). A few mammalian remains have been found in Cretaceous and Jurassic rocks, but it is true to say that there is no evidence of the existence of the large and modern type of mammal before the beginning of the Tertiary era. The Cretaceous period belongs to a geological era which has been called the Age of Reptiles, animals such as lizards, snakes, crocodiles, and other egg-laying, cold-blooded creatures. Crocodiles, turtles and other reptiles lived in the water and on the muddy flats of the Wealden lake, but those that were then the most conspicuous members of the class have long been extinct; they were Dinosaurs. Some Dinosaurs greatly surpassed in size even the largest mammals and were very much larger than any existing reptile. In 1822, bones of one of these extinct beasts were discovered in the Weald district at Uckfield, and the animal was called *Iguanodon* from a resemblance to the South American *Iguana*: other remains have been found in more recent years, particularly an important discovery made by an amateur, R. W. Hooley, a business man in Winchester who was able to put together a complete skeleton from bones collected in beds of Wealden age at Atherfield in the Isle of Wight. *Iguanodon* was a vegetarian; it walked on long hind legs and used its shorter fore-limbs to reach boughs of trees; it was 15 ft. or more in height. Many other Dinosaurs have been discovered in Cretaceous rocks, especially in North America and tropical Africa. This strange race of animals died out before the Tertiary era; their huge unwieldy bodies and extremely small brains seriously handicapped them in the struggle for existence. There were also flying reptiles (*Pterodactyl*) provided with large wings attached to the elongated fifth finger of each hand. One of the best known extinct reptiles is *Ichthyosaurus*, discovered by Miss Mary Anning, in 1828, in Jurassic rocks near Lyme Regis; this beast, characterized by large swimming paddles and a long snout, together with the stream-lined, long-necked and smaller-headed *Plesiosaurus*, lived in Cretaceous and Jurassic seas.

The single word Cretaceous, used in a geological sense, connotes a period of earth-history the significance of which is beyond our power to grasp; it covers a stretch of time that we can only feebly describe as immeasurable. The little we know, and it is very little, of the changing face of the earth and the procession of living things that moved across the world's stage creates a desire to know more of an age when the old order changed, giving place to the new. We should like to know much more of the life on the continents washed by Cretaceous seas and gulfs and on islands set in the waste of waters. It was on the land, as well as in the seas, that great transformations were made; the Reptilian dynasty was being gradually ousted by the rapidly growing strength of the dynasty of Mammals. The scraps of vegetation embedded in the sediments from seas and lakes give only a tantalizingly imperfect picture of the forests and undergrowth in regions beyond the reach of rivers and other transporting agents; in the Cretaceous period the vegetation of the world assumed its present dress and replaced an older one that had previously held sway over the whole surface of the land. It was in the course of the Cretaceous period that the animal and plant world passed from an ancient to the modern type; it was an age of transition, an age in which the stage was set for a new company of actors, excepting only man, whose time was not to come until millions of years had passed.

CHAPTER XI

Estuaries, Lakes and Seas from the cliffs of Yorkshire to the coast of Dorset

Penetrating below the sandstones, clays and other rocks deposited as sediment on the floor of the Wealden lake on the surface of which were reflected the trees and undergrowth in the early stages of the Cretaceous period, we reach a thick pile of other sedimentary rocks containing records of the next older period in the history of the earth that was long ago named by a French geologist the Jurassic age. The name was taken from the Jura Mountains on the frontiers of France and Switzerland which are made of regularly folded rocks upraised from the bed of a Jurassic sea. It was on fossils collected from rocks of the same age in south-west England that the Father of English geology, William Smith (see Chapter II), based his epoch-making statement: 'The same strata are found always in the same order of superposition and contain the same peculiar fossils.' Names given by him to successive sets of rocks included in the Jurassic period are still in use, but for our present purpose it is unnecessary to do more than refer by name to a few of the subdivisions. The rocks of this period make up a pile of sandstones, clays and limestones having in some places a thickness of many hundred feet and representing in time about 25 or 30 million years, rather less than half the duration of the Cretaceous period. It will be convenient to commit to memory a few of the names given to the more important series of sedimentary beds comprised within the period. The youngest Jurassic rocks are spoken of as Upper Jurassic and include in descending order the Purbeckian and Portlandian series which are displayed in their greatest development in the Isle of Purbeck and the Isle of Portland respectively; next below them is the Kimmeridge Clay, which is preceded by rocks of another kind known as the Corallian because of the abundance of fossil corals; below this series is another thickness of clay known as the Oxford Clay. Below the Oxford Clay is the Great Oolite series, followed at a lower level by the Inferior Oolite series: as the name implies, many of the rocks of these series have an oolitic structure (see Chapter IV, p. 49). The oldest series in the Jurassic period is the Liassic stage.

As we follow the changing face of the earth through the aeons of geological history and trace the events dimly recorded in the structure of the rocks and in the fossils many of them contain, we discover evidence of alternating periods of far-reaching and intense upheavals and folding of the crust accompanied by protracted volcanic activity, followed by long interludes of relative stability when the raw material of the rocks of a later age was quietly accumulating on the floor of oceans and fresh-water lakes. Throughout the long Jurassic age the rocky founda-tion of the British area, like other parts of the world, was not subjected to any great crustal disturbance; there was no flooding of the land by lava-flows, no building of mountain chains, but an almost uninterrupted deposition of water-borne sediment and the slow growth of limy ooze consisting of shells and other hard parts of marine animals on the floor of a clear and relatively deep sea. The distribution of marine, brackish-water and fresh-water sedimentary rocks in the English area is proof of the incursion of Jurassic waters over much of the country that is now land. Jurassic rocks form a fairly broad band from Redcar, in county Durham, through East Yorkshire, Lincolnshire, Rut-land, Northamptonshire, Cambridgeshire, Bedfordshire, and Oxfordshire and on to Portland Bill on the Dorset coast. During the long period the whole of the surface where the rocks are exposed was at one time or another either an open, a land-locked sea, a delta or an estuary. In addition to the exposures over the continuous strip of England from the county of Durham to Portland Bill there are other places in England and Scotland where rocks of the same period occur, either as isolated patches or as detached hills not far from the main mass. There are exposures of Jurassic rocks in the Carlisle district, between Brora and Helmsdale on the north-east coast of Scotland, in some of the Inner Hebrides and in north-east Ireland. Glastonbury Tor and Bredon Hill in Gloucestershire are what are called outliers, which means outlying portions left by the erosion of rocks which originally linked them with the main Jurassic belt. It is highly probable that the Jurassic rocks in Scotland are remnants of widespread sheets formerly covering a much larger area. When a considerable part of the British Isles was under water there was a large land area on the site of the North Atlantic Ocean; the mountains of Wales, parts of Devon, Cornwall and Brittany were islands near the edge of the ancient continent. All the Jurassic rocks are sedimentary in origin; none are volcanic. The most characteristic feature is the variation in the nature and manner of formation of many of the sediments as they are traced from one district to another; in some localities, e.g. in the Cotswold Hills, most of the rocks are limestones uplifted from the bed of a clear sea; in others, sandstones and clays afford evidence of shallower water either near the sea-shore or in the broad estuary of a river. Beds with marine fossils in one district are represented in another by contemporary beds containing remains of estuarine animals and samples of land vegetation. This variation is apparent not only as the Jurassic rocks are traced laterally over the whole belt, but there is a similar variation when the beds are followed from layer to layer in a vertical direction. The changing character of the sediments is indicative of rising and falling of the surface, of a rhythmic oscillation in depth of the water and in the physical environment.

Immediately below the oldest Cretaceous rocks we come to the youngest Jurassic sediments which are seen in their most impressive development in the Isles of Purbeck and Portland. Sediments of the same age occur at Shotover Hill near Oxford, at Swindon, and at many other places. To the Purbeck and Portland series we owe some of the grandest coastal scenery in southern England, the cliffs of Durlston Head south of Swanage and those farther west at St Albans Head. The perpendicular cliffs of Portland Bill are composed mainly of limestone containing numerous marine shells. The Purbeck and Portland rocks, including limestone, sandstone and other sedimentary beds, form isolated hills and winding ridges overlooking low ground occupied by softer and less resistant Jurassic and Cretaceous clays. The youngest of these sedimentary beds, the Purbeckian series, mark a transition from the marine rocks of the Portland series to the fresh-water beds of the early Cretaceous Wealden lake. In the latest or Purbeckian phase of the Jurassic period the southern district of England was a land of fresh-water lakes and swamps: this we know from the fossil content of the rocks. On the cliff east of Lulworth Cove one sees trunks of trees encased in a chalky substance that was probably precipitated over them as they lay in a pool of water saturated with carbonate of lime. These fossil plants and the associated material are an example of a land-surface. Large, partially petrified stems, many feet in length, of extinct cone-bearing trees are common objects in the Isle of Portland. A well-known rock from the uppermost Jurassic series is the Purbeck marble; it is a mottled green and red limestone called marble because it can easily be polished; in it are numerous shells of a fresh-water snail known as *Viviparus*, and formerly called *Paludina*. Purbeck marble was extensively used in the thirteenth century for shafts and pillars in mediaeval cathedrals: good examples can be seen in Salisbury Cathedral, York Minster and many other Gothic buildings. The important point is that the geographical and physical conditions in southern England at the time of the Wealden lake were preceded in the closing stage of the Jurassic period by very similar conditions; there was no clearly marked break in continuity. The Portland rocks underlying those of Purbeck age are marine in origin, rich in ammonites and the shells of many other creatures, and are proof of an area of sea which was later occupied by swamps and lakes. The steep Dorset cliffs are made of limestone that was originally a chalky ooze on the floor of the Portlandian sea. Portland stone was first used by Inigo Jones, in 1619, in the construction of the Banqueting Hall of Whitehall, and by Sir Christopher Wren in the rebuilding of St Paul's and for many London churches: it is a light grey stone which turns whiter after exposure to the weather. If one looks closely at Portland stone, one can readily detect pieces of many kinds of shells of animals which contributed more than 100 million years ago to the formation of one of the best building stones in the world.

Continuing our descent below the Portland series, we come to a thick mass of clay, the Kimmeridge Clay, named after Kimmeridge in Dorset where, as in many other parts of England, it is well developed. The same clay has been traced from Yorkshire, through the Midlands, to southern England; it contains marine shells, bones of fishes, and other animals, and was deposited over a large area on the floor of a muddy and relatively shallow sea. Kimmeridge Clay underlies many flat tracts of country including some of the Fenland. Below the clay there follows a series of rocks that are spoken of as Corallian because of the occurrence of many fossil corals that are remains of reefs in a clear sea. Rocks of Corallian age, exposed near the village of Upware, 10 miles north-east of Cambridge, mark the position of coral reefs in the Jurassic sea: other exposures are seen in the shores of Weymouth Bay, at Filey Brig, and in the Castle Rock at Scarborough. The next stage in descending order is represented by a second thick mass of clay known as Oxford Clay; it underlies the city of Oxford and occurs at many places in southern England, also as far north as Lincolnshire and Yorkshire; it forms the foundation of much of the East Anglian Fenland and is responsible for the monotonous scenery which has its compensation in the wide expanse of sky, the sunsets and a certain sentimental attractiveness felt by people who have lived in the Fen country. When the Oxford Clay was slowly increasing in thickness as mud on the sea-floor, the Jurassic sea lay between two land-masses and was joined to a wider sea by a strait bounded on the west by the Welsh hills that formed the eastern edge of the Atlantic continent. From the numerous Oxford Clay pits in the Peterborough district with groups of tall chimneys of brickworks the brothers C. E. and A. N. Leeds, notable examples of amateurs who have substantially contributed to knowledge of ancient life, made a remarkable collection of skeletons of Ichthyosaurs and many other extinct reptiles that abounded in Jurassic seas. Oxford Clay lies below the Vale of the White Horse, in the valley of the Evenlode, in Bedfordshire and Huntingdonshire and along the borders of the Thames valley. Beneath this thick deposit of clay, belonging to an earlier stage in the period, there are sedimentary beds of limestone and others of shallower water origin: layers of rock assigned by geologists to the Great Oolite series are seen in the L.M.S. railway cutting at Roade and other places; the Bradford Clay, named after Bradford-on-Avon not far from Bath, is of this age. The rocks seen in the embankment at Roade are mentioned as one of many examples of interesting exposures of beds of different ages that one sees in the course of railway journeys; their age can generally be made out by reference to the geological map of the British Isles. The well-known Stonesfield roofing slates (mentioned in Chapter IV) are made of sandy limestone, and similar thin beds occur in the Cotswold country. The Stonesfield beds are famous as the source of some of the oldest known fossils of mammals. One of the best districts for the study of Jurassic rocks in England, especially limestones belonging to the Great Oolite and the Inferior Oolite series, is the Cotswold country; most of the Cotswold Hills are in Gloucestershire, and are part of the long Jurassic belt; the steep western escarpments of relatively hard rock overlook the Vale of Evesham and Gloucester: towards the east the surfaces of the escarpments slope more gently. The whole district is a country of rolling plateaux and wolds of limestone with well-wooded narrow valleys. The local grey-brown stone and the roofing slates of the attractive Cotswold buildings are from sediments upraised from a Jurassic sea. Much of the stone is made of debris from coral reefs battered by the waves under conditions comparable to those in the Great Barrier Reef, off the eastern seaboard of Australia. Rocks of Middle Jurassic age, many of them oolitic in structure, have long been used as building stones: Barnack stone, reputed to have been quarried thirteen centuries ago, was employed in the construction of Ely Cathedral, of Peterborough Abbey and in churches at Bury St Edmunds. The darker columns in Ely Cathedral are Purbeck marble and the Lady Chapel arcading is carved in a homogeneous hard Chalk. Other well-known stones are from Ancaster in Lincolnshire, Ketton in Rutlandshire, and other places. The yellow-brown and pink Ketton stone is one of the best examples; its tightly packed oolitic grains are a striking feature; when examined in a thin transparent section one can see that each grain consists of concentric layers of carbonate of lime with a central nucleus such as a small fragment of a shell. The oolitic structure is usually supposed to be due to the gradual deposition of concentric layers in shallow water rich in lime as the particles were rolled to and fro by currents. It is probable that this structure, characteristic of many Jurassic limestones, is the result in part of the action of very small and simple water-plants which caused the precipitation of carbonate of lime. Modern oolite occurs on the shores of the Great Salt Lake of Utah and the Red Sea. In addition to the building stones already mentioned, the Bath stone, included in the Great Oolite series, is a good example of oolite; it was used by the Romans in their buildings at Bath, by the mediaeval masons at Glastonbury Abbey, and in many later buildings. Most of the Oxford colleges and University buildings are built of Jurassic oolitic rocks from quarries at Headington, Wheatley, Cowley, and other places: many of the roofs are covered with Stonesfield slate. Some of the buildings afford grotesque examples of the destructive action of the weather, notably the gigantic heads on tall pedestals surrounding part of Wren's Sheldonian theatre. Ketton stone is seen to great advantage in Wren's library at Trinity College, Cambridge, also in the old and new buildings of Downing College.

Wells Cathedral is built of rock known to geologists as Inferior Oolite, a term apt to be misunderstood: a Cambridge under-graduate asked his tutor, who was a geologist, what stone had been used in the construction of the college chapel and, on being told it was Inferior Oolite, replied: 'Just like the authorities to use inferior material.' The ridge of rock running in a north–south direction in Lincolnshire, known as the Cliff, on which stands Lincoln Cathedral and along whose summit runs the Roman Ermine Street, is mainly Inferior Oolite limestone, and some of the Cotswold rocks are of the same age. Some of the rocks included in the Inferior Oolite series are marine in origin; others were deposited in estuaries and deltas. When we follow the Inferior Oolite series to the East Yorkshire moorland

and the high cliffs on the coast, we find a thick series of sand-stones, shales and other sediments, many of which contain plant remains and animals that lived in deltas. The land that is now East Yorkshire is formed, in part, of deltaic rocks and, in part, of rocks deposited on the floor of an open sea. When the more southern and south-western districts were under the sea, broad deltas lay over the Yorkshire area to which rivers transported sand and mud, and a varied assortment of land plants lived on the mud flats as well as on the neighbouring land.

Passing farther down to rocks below the marine oolites and the estuarine sediments we reach the oldest series of the Jurassic period, called by geologists Liassic; the term Lias is said to have been used by quarrymen for layers of rock. Liassic rocks include sandstone, shale, and limestone containing many shells of molluscs, sea-lilies, skeletons of *Ichthyosaurus*, *Plesiosaurus* and other reptiles of the sea, and in some places, e.g. near Lyme Regis, well-preserved land plants. Rocks of this age are well displayed in the face of the cliffs from Bridport to Lyme Regis and Seaton; in the cliffs west of Lyme Regis the alternation of thin beds of limestone and shale illustrates changing conditions in the sea-floor, a rhythmic up-and-down movement from clear water to water containing sand and muddy sediment. Liassic rocks occur in the Vale of Evesham, at Melton Mowbray in Leicestershire, near Towcester and at many other places. Clays of the same Jurassic stage occur along the western border of the Cotswold Hills, from near Tewkesbury to Bath, on the edge of the Mendip Hills and in the cliffs at Watchet on the Severn estuary. Farther north rocks of the same age reach a considerable thickness on the Yorkshire coast at Robin Hood's Bay, in the cliffs at Peak over 600 ft. high, and other places where sandstones and shales take the place of the deeper water limestones in more southern districts. Some of the older Jurassic rocks are valuable sources of iron ore, in the Cleveland Hills and at many places in Northamptonshire and the Kettering district rocks with oolitic structure consist mainly of iron compounds. At almost all states of the tide at Whitby and farther south one sees a line of breakers parallel to the coast caused by the existence of a shelf of what is known as the Jet Rock, because it was from this rock in the neighbourhood of Whitby that jet used to be obtained. Jet consists of much altered pieces of wood in which little structure can be seen; some of it was originally wood of an *Araucaria*. The Jet Rock against which the sea makes a line of breakers is the truncated edge of a platform running west from the foot of Saltwick Nab as a sunken reef and ending just off the East Cliff, crowned with the ruins of St Hilda's Abbey, at the mouth of the Whitby estuary. A comparison of the Jurassic rocks on the east and west sides of the harbour at Whitby shows that the beds on the two sides do not lie at the same stratigraphical level; those on the west side lie about 200 ft. lower than the corresponding beds on the east side. Originally the rocks were continuous across the estuary, and their present lack of correspondence is evidence of a line of fracture—a fault —which no doubt was the initial cause of the present position of Whitby harbour. The striking contrast between the Liassic rocks of Yorkshire and Dorset is an expression of differences in the circumstances of their deposition. There was a clear and deeper sea in the south and in the north shallower water of deltas or river estuaries. During the early stage of the Jurassic period the greater part of England was covered by sea; most of Wales was land, parts of Devon and Cornwall, and the Mendip Hills formed groups of islands. Part of Scotland was an island in the Liassic sea, sediments of which are still left as remnants of larger sheets in some of the Inner Hebrides and on the east coast of Sutherland.

The salient features of the Jurassic period may be summarized in general terms: the uppermost sedimentary rocks—those known as Purbeckian—which lie immediately below those included in the oldest subdivision of the succeeding Cretaceous period bear witness in their fossil content to the existence of fresh-water conditions and to continuity between the Wealden lake and the broad tract of fresh water which left records of its existence in the upraised sediments in the Isle of Purbeck. Below the Purbeck beds there is a very thick pile of older Jurassic rocks varying much in the character of the sediments and including all the subdivisions from the Portland to the Liassic series. A striking feature of the Jurassic rocks as a whole, when they are traced from south to north, is the recurring alternation of limestones, clays, and sandstones: this alternation is shown when the rocks are examined in a vertical direction, that is, through a succession of beds lying one above the other in one district in the order of age; it is also apparent when beds of the same age are traced laterally from place to place from south-western to north-eastern England. The Kimmeridge and Oxford clays are examples of deposits of the same kind but belonging to different periods of time which afford evidence of turbid sea water. Contrasts are revealed by a comparison of sedimentary rocks in the Cotswold Hills and other districts in the south-west of England with beds of the same age in East Yorkshire. Limestones in the south with abundant marine shells and corals, when traced to other districts, are replaced by sediments of shallow-water origin and in places by sandstones penetrated by roots of plants that grew on land or swampy soil. These contrasts in the nature of the rocks and of the fossils enable us to visualize differences in the landscape: an open sea with a range of hills forming the border of a continent to the west; farther north a great delta with a sparse covering of plants, and beyond that woodland on the banks of rivers.

It has been possible to form a general idea of Jurassic vegetation: some of the oldest rocks containing plants are on the coast of Dorset, plant-bearing beds belonging to higher stages occur in Oxfordshire and neighbouring counties, and the richest of all in Yorkshire, in the cliffs at Whitby, Hayburn Wyke, and near Scarborough, beds of shale at Gristhorpe and Cayton Bay, and at inland localities such as Roseberry Topping and Marske. There are differences in detail between the Jurassic floras reconstructed from collections made at different stages of the period, but throughout the millions of years covered by the whole period the vegetation, in its main features, was fairly uniform in character. There was little contrast between the plant world in the early stage of the Cretaceous period, as illustrated by the fossils collected from rocks of Wealden age, and that which clothed the borders of the Jurassic continent. The expression 'borders of the Jurassic continent', is used deliberately because we must remember that our knowledge of the vegetation is almost entirely limited to the remains of plants that lived within reach of rivers and their tributary streams and had, therefore, a chance of being preserved in the water-borne sediments. The most striking and convincing evidence of the wide distribution of closely related Jurassic plants and the astonishing similarity in the vegetation of places separated from one another by many thousand miles was furnished by the discovery, some years ago, by members of a Swedish expedition of a large number of fossils, chiefly leaves and branches of trees and ferns, in Graham Land, lat. 63° 15′ S. on the edge of the Antarctic continent. Several of the plants proved to be very closely related to, if not identical with, species previously recorded from England and India. It would be an exaggeration to describe the Jurassic vegetation of the world as uniform in the strict sense; there were local differences, but they seem to have been slight and relatively insignificant in comparison with the amazing resemblance of the floras of the same geological age which left samples in places remote from one another and situated in many different latitudes. Some of the Graham Land plants differ very little from species which lived in Greenland in the early days of the Cretaceous period. These facts have a special importance and interest because they illustrate a problem which often crops up when fossils from widely separated regions are compared, namely the difficult and perplexing problem—how to explain the very great differences between climatic conditions in the past and those of the present. It is not only fossil plants which bring us face to face with this question of past climates; fossil animals of many geological periods tell the same story. As in the earliest Cretaceous flora so also in the Jurassic floras there is practically no evidence of the existence of flowering plants, the class which by the latter part of the Cretaceous period had risen to an important position and wandered over most of the earth's surface. The only fossil of Jurassie age which is in all probability the leaf of a broad-leaved tree was found several years ago at Stonesfield, in Oxfordshire. There can be little doubt that plants existed in the Jurassic period closely related to modern flowering plants, but, as yet, we have very little knowledge of them. Specimens were described, some years ago, from the Yorkshire coast south of Scarborough which, it was thought, might be forerunners of the flowering plants chiefly because they possessed fruits containing seeds, but it was subsequently found that the wall of the fruit did not completely enclose the seeds as it does in flowering plants. It is, however, possible that these extinct fossils illustrate a line of evolution which ultimately led to the true flowering plants. Leaving this difficult and still unsolved problem of the origin of the present dominant class in the plant kingdom, let us look at a few of the commoner members of Jurassic floras. There were many cone-bearing trees related to existing species. There were British araucarias, representatives of a genus now unknown as a wild tree in the Northern Hemisphere, cypresses and allied trees related to the cypresses that are now characteristic of southern Europe and Asia Minor. The 'living fossil' *Ginkgo biloba*, the maidenhair tree, had ancestors in the Jurassic forests not only in western Europe but almost all over the world. In the Jurassic as well as in the early part of the Cretaceous period one of the most characteristic and widely distributed classes of plants was an extinct class usually spoken of as the Cycads. Cycads are naked-seeded plants now almost entirely tropical; they are more fully described in Chapter XVII. Many of the so-called Jurassic Cycads differ widely in their reproductive organs from any living plant and belong to an extinct group; some, on the other hand, were undoubtedly related to existing Cycads. The important point is that in the Jurassic and early Cretaceous woodlands plants belonging to a group that has long ceased to exist played a prominent part. There were horsetails resembling, except in their much more robust and larger stems, the common horsetails (*Equisetum*) still living in Britain; there were also some relatively small plants closely related to our club-mosses. Ferns were well represented: species closely related to our Royal fern (*Osmunda regalis*), and others whose nearest living allies are no longer found in Europe but in the southern tropics and sub-tropics. The Jurassic vegetation, though separated from that of the age in which we are living, by more than 100 million years, was in some respects not very different from the plant-world of to-day: the greatest contrast is the apparent absence of flowering plants—no broad-leaved trees such as oaks, elms, beeches, limes and many other familiar kinds.

Jurassic seas were inhabited by a varied population of marine creatures, many of which are still represented by descendants living in our seas of the present age; other animals that were among the most abundant, such as ammonites—the snake-stones of St Hilda—belong to groups that have long ago ceased to exist. A characteristic shell common in some Jurassic limestones

—the genus *Trigonia*—agrees very closely with that of a mollusc which lives in Australian seas. The greatest contrast between the Jurassic and the modern animal world is in the higher animals; there were no birds like those of to-day, only strange creatures like *Archaeopteryx*, the oldest known animal provided with feathers and, unlike living birds, with teeth. This extinct genus was in some respects more like reptiles than birds and furnishes evidence of the probable descent of birds from reptilian ancestors. Large reptiles of the sea, closely related to those found in Cretaceous rocks, including *Ichthyosaurus*, *Plesiosaurus*, and several others were abundant and widely distributed: the Jurassic period was an age of reptiles, many of which were gigantic and bizarre products of evolution, Nature's misfits that endured but for a season. In the course of our backward journey we have noticed the decrease in mammals, both in numbers and variety as well as in size. The highest class in the animal kingdom had not yet played a prominent part; a few fragmentary remains of small mammals have been found in Jurassic rocks, and it is interesting to note that they have little in common with any living groups of mammals. On the other hand the remains of certain distinctly mammal-like reptiles found in pre-Jurassic beds definitely suggest that the mammals, like the birds, had a reptilian ancestry.

From a knowledge of the animal and plant worlds of other days we are able, with more or less confidence, to draw a distinction between ancient and modern members of the two kingdoms of our own day; fossils help us to recognize among the living links with the past, survivals that have persisted while other groups of animals and plants flourished for a comparatively short time and for reasons at which we can only guess ceased to be.

CHAPTER XII

Salt Lakes and Deserts

There is no well-defined boundary separating the rocks referred by geologists to the initial stage of the Jurassic period from those which it is customary to include in the Rhaetic period, and yet it is abundantly clear that the rocks underlying the upraised sediments of the Liassic sea bear witness to a change in physical conditions. It is difficult to say exactly at what precise level in the sequence of sedimentary beds the change occurred; it was a gradual change which becomes more clearly defined as we continue our backward journey. We have reached another transitional epoch comparable with that between the fresh-water Wealden rocks at the beginning of the Cretaceous period and the marine limestones of the Upper Jurassic or Portlandian sea; but this earlier transition is of a different kind. As we have seen, the oldest rocks of the Jurassic period contain a varied and abundant assemblage of the remains of animals characteristic of a clear, open sea which covered the greater part of England. Next below them are sedimentary beds poor in fossils and containing a few shells that appear to be stunted and abnormal in striking contrast to those in the overlying Liassic rocks. Moreover, in the strata below the Rhaetic sediments are layers of gypsum (sulphate of calcium), a sign of land-locked seas or lagoons in a dry or semi-desert country where lack of rain caused precipitation from water saturated with salts, as in the Caspian and the Dead Sea to-day. The name Rhaetic was taken from the old Roman province of Rhaetia, between the basins of the rivers Po and Danube, where rocks of this period are well developed. Rhaetic rocks in England are represented by a thin series of sedimentary beds mostly barren and at their greatest thickness not more than 100 ft. The sediments included in the Rhaetic period are classed by some geologists as Jurassic and placed below the marine beds of Liassic age; by others they are assigned to the uppermost part of the Triassic period. They have often been described as passage or transition beds linking the Triassic and Jurassic periods, but the very great thickness of Rhaetic rocks in the European Alps is evidence that they represent a long period of time, much more than a mere transitional stage. In the British area we have only a few pages of this chapter of geological history, only a few episodes which have left thin sediments and a small number of animals and plants. An account of a remarkable Rhaetic plant (*Naiadita*) is given in Chapter VI, one of the few found in this country in contrast to the unusually rich collections yielded by rocks of the same period in other parts of the world. They are exposed at several localities along a belt of country stretching from Redcar, on the north-east coast of Yorkshire, diagonally across England to the south and south-west; they are well displayed at Penarth, near Cardiff, at Watchet, on the Bristol Channel, at Westbury-on-Severn and many other localities. The uppermost and youngest beds are limestones which occur at Cotham, north of Bristol, known as the Cotham or Landscape Marble because of black markings on its surface which resemble miniature trees a few inches in height and were erroneously described as fossil mosses; they are inorganic deposits of mineral—a black oxide of manganese. This limestone, usually not exceeding 20 ft. in thickness, occurs at many places: it is sometimes called White Lias; it differs from the typical Liassic limestones in being thinner and in containing very few fossils, including stunted shells of an extinct oyster. One of the most interesting Rhaetic rocks is a thin layer known as the Bone-bed; it is a black shale (mud) with many bones of fishes and reptiles, and is exposed at localities in Leicestershire, in the Severn valley and elsewhere, e.g. at Watchet and Westbury-on-Severn. One of the fishes (*Ceratodus*) is an extinct type related to the primitive lung-fish still living in Australian waters. The large accumulation of bones suggests the destruction of life by a sudden overflow of the sea-bed by foetid mud. A comparison has been made with deposits of black mud now being formed on the bed of the Black Sea, which is near a desert region with low rainfall and a high summer temperature. The Black Sea is poor in life; in it salt water forms a stagnant pool and bones of fishes fall into the foetid mud where the salt water preserves them.

At the beginning of the Rhaetic period a considerable part of England was submerged under a sea in which animal life was poorly developed; a sea similar to the Caspian and the Dead Sea, in which gypsum and other salts were deposited. In the Rhaetic sea *Ichthyosaurus*, *Plesiosaurus*, and other reptiles were able to exist, but we have now reached the lowest level in the geological series at which their bones have been found. Among other animals in the Bone-bed are a few small mammals, the earliest examples of the class that was destined to play a leading part in later periods: these oldest known mammals were no larger than rats or mice and are quite unrelated to any living types. The conditions reflected in the Rhaetic rocks are the epilogue to more widespread deserts and salt lakes in the longer Triassic period to which we now pass. The name Triassic, first used by German geologists in 1834, implies a threefold grouping, a division of the rocks of the period into three well-marked series, called respectively Keuper, Muschelkalk, and Bunter. In the British Isles the middle series is not developed, and we need not concern ourselves with it. The word Keuper was a local term in use in Coburg, and Bunter, from a German word which means variegated, has reference to the parti-coloured sedimentary beds.

The uppermost beds of the Triassic period show that over a large proportion of the land that is now England there was typical desert scenery, wide stretches of sand, salt lakes and a relatively high temperature. In many parts of the country from the Eden valley to southern England the red colour of the fields, the red rocks in railway cuttings and cliffs is, in most districts, evidence of the occurrence of Triassic rocks, but colour alone is not a trustworthy clue to geological age. There are other and older rocks that are red, sandstones of Permian age, and the still more ancient rocks of the Old Red Sandstone. It is, however, true that the majority of red rocks in England are either Triassic or Permian, and sometimes the two together are spoken of as New Red Sandstone, to distinguish them from the Old Red Sandstone that plays a prominent part in Scotland and the islands to the north of the Pentland Firth. The Keuper series consists largely of parti-coloured muds and clay, beds of red sandstone, with intercalated layers of rock salt, gypsum, and alabaster. Salt is obtained on a large scale in Cheshire, Worcestershire and in the Middlesbrough district: such place-names as Nantwich, Northwich and Droitwich owe their termination to a Saxon word which means a Wych or salt-house. On the surface of Triassic sediments there can occasionally be seen large footprints of extinct amphibian animals that walked over the soft, exposed, sandy and muddy flats bordering a salt sea, footprints that are sometimes the only records of creatures whose bones perished. There are also cracks in Keuper beds that form an irregular network and were made by contraction, on drying, of a surface warmed by the sun; pits made by the impact of rain where rare rain-storms spattered the arid country, and wavy ripple-marks. Another indication of the circumstances in which Triassic rocks were deposited is the frequent occurrence of current-bedding often well displayed on the weathered faces of sandstone, in which the action of the weather

PLATE VII

Photograph by P. W. Wright

Triassic desert landscape

Granitic rocks in Charnwood Forest, grooved and polished by Triassic sandstorms

has etched with relief series of much narrower layers lying at varying angles to the major planes, indicative of shifting currents in shallow water. All these features help us to reconstruct the conditions under which the sedimentary beds were formed.

One of the convincing and impressive proofs of a desert landscape is furnished by the granitic rocks of Charnwood Forest, in Leicestershire (Plate VII). These rocks were not completely covered by the Keuper sediments deposited on the floor of a salt lake; the higher ground was left as an island, but the lower slopes of the hill became encased in layers of Triassic clay and mud. The relatively soft covering was an easy prey to denudation and much of it was removed, leaving exposed the face of a cliff that had long been hidden. The exposed rocks provide part of a landscape revealed by eroding agents which wore away the mantle of sediment that had been spread over them more than 100 million years ago. The Charnwood granite face is smooth and polished, and across it are horizontal, broad, rounded grooves; it is an exact counterpart of a rocky escarpment in a modern desert where strong winds hurl blasts of sand from dunes and the desert surface against rock faces, polishing them, etching them and scouring out grooved channels, 'it was a barren and dry land where no water is'. Other evidence is furnished by the layers of salts deposited on the bed of a Triassic Dead Sea as the hot sun evaporated the saline water; on the surface of some Keuper sediments one can sometimes see small rectangular blocks of clay or sand projecting from the rock: these are pseudomorphs of crystals of rock salt, that is, crystals that were replaced as they dissolved by the surrounding sedimentary material. The water used in breweries at Burton-on-Trent comes from Keuper marls and it is said to impart a certain quality to Burton ale by reason of the salts which it contains. The cathedrals of Chester and Hereford are two of many large buildings made of red Keuper sandstone. On the flanks of the

Mendip Hills Triassic beds abut on the slopes of higher ground formed of older rocks, as in Charnwood Forest. The Quantock Hills in Somerset were also islands in the Keuper sea.

Below the Keuper series, in the British Isles, we come to the lowest sedimentary beds of the Triassic period, the Bunter series, which includes, in addition to clays and marls, red sandstones and thick masses of pebble-beds or conglomerates. The pebble-beds are the most impressive rocks of the series; they occur over a wide area from Doncaster to Birmingham, in Staffordshire and Cheshire, and are especially well developed in the cliff at Budleigh Salterton on the Devon coast; they often form high ground in contrast to the low-lying plains on the much softer Triassic beds. Nottingham Castle stands on a hill of coarse Bunter sandstone and pebble-beds. The Budleigh Salterton cliff consists in the lower part of a thick mass of rounded pebbles embedded in a red matrix overlain by sandstone. Standing on the shingle beach and looking at the pebbly cliff rising above the present shore-line, we realize the striking similarity of the loose shingle below our feet and the cemented mass of pebbles in the cliff, two shingle deposits separated in time by well over 150 million years. Listening to the swish of receding waves and the music of the pebbles, one can hear an echo of the same sounds on the fringe of a Triassic sea. In addition to rounded and water-rolled pebbles the coarse red rocks of the Bunter series also contain angular stones. There has been much discussion over the provenance of the stones, many of which are made of a very hard quartz rock, a sort of indurated sandstone, and fortunately some of them contain fossils well enough preserved to provide a clue to the geological age of the parent rock. The general view is that one source of the pebbles was a much older rock in northern France, from which they were carried by rivers flowing across part of an ancient continent that extended from France to southern England. The large size of the stones and the well-rounded form of many of them show that they must have been carried by water in spate, torrential streams rushing down hill slopes, and spread as fans over a flat desert plain. Comparable fans of detritus are characteristic of dry mountainous countries at the present day, where, from time to time, long spells of drought are followed by torrential rains which fill the depleted watercourses and carry with irresistible force a heavy load of stones. The more angular stones, that are a conspicuous feature in some of the Bunter beds, were, no doubt, transported a shorter distance and derived from piles of broken rock detached by the action of the weather from cliff faces such as we can see in the screes of Wastwater and other places in hilly districts. These Bunter pebble-beds and the finer grained sandstones are all made of material carried for longer or shorter distances from parent rocks bordering inland salt seas and lakes. The Bunter sands and pebble-beds, reddened by films of an oxide of iron over the individual grains and pebbles, take us back through the ages to an English landscape reproduced in our own day in many desert lands where there are ranges of sand-dunes, sharp-edged and ripple-marked by the wind, lakes heavy with salt, and recurrent rain-storms, hissing torrents laden with the products of destruction, carrying to lower levels loose stones from sand-blasted hills.

The distribution of land and water over that part of Europe that is now the British Isles bore little resemblance to the present geographical features: most of Scotland was land which was continuous across Ireland and beyond into the Atlantic Ocean. The Highlands of Scotland and the Welsh hills were then, as now, mountain ranges; part of the Pennine Chain was an island in the Triassic sea. Farther south, the hills of Charnwood Forest, the Mendips, Dartmoor and Exmoor and the headlands of Cornwall were dominating features in a dry, Dead Sea land. The British scene is typical of much of northern Europe and the eastern region of North America, where red rocks similar to those in this country are the foundation-stones over a large area: desert and semi-desert conditions prevailed over a large region of the Northern Hemisphere. When we extend our survey of Triassic rocks to southern Europe a very different story is unfolded: enormous masses of limestone, rich in marine fossils, corals and many kinds of lime-secreting sea-weeds, afford convincing evidence of a broad sea and genial climate. The glorious Dolomite Mountains of the Tyrol have been sculptured out of upraised calcareous sediment that, in the course of millions of years, grew in thickness on the bed of the Tethys Sea which lay across Europe and Asia. Desert waters, dunes and salt lakes to the north; in the south the Tethys Ocean, with reefs of coral and banks of coral-like seaweed: such were the contrasts in the Triassic world.

It is convenient to include in this chapter a short account of rocks which are assigned to the Permian period and regarded as marking the upper limit of the Palaeozoic era, the Triassic period being usually spoken of as the first or oldest stage of the Mesozoic era. For our present purpose it makes for greater continuity in the backward journey to draw the line below, and not above, the Permian period. The name Permian was given to this period of geological history—lasting, it is believed, nearly 40 million years, longer than the Triassic or the Jurassic periods—more than a century ago because of the great development of the rocks on the flanks of the Ural Mountains in the Russian province of Perm. Rocks of Permian age are red sand-stone marls, conglomerates, cream-coloured limestones and other sediments: many of them are very similar to Triassic rocks both in appearance and in the evidence they furnish of arid conditions. They occur on both sides of the Pennine Chain; at Penrith, in Cumberland, where one sees red rocks close to the railway station; in the Eden valley, in the Manchester district and many other places. East of the Pennines, Permian rocks occur in Durham, at Darlington and Ferryhill, in the cliffs at the mouth of the Tyne and along the coast near Sunder-land, where the cliffs above the sandy beach are built of creamy limestone eroded into detached stacks. Rocks of this age are exposed over a narrow band of country from the Durham coast to Nottingham: the dripping well at Knaresborough, in the valley of the Nidd, comes from the foot of a cliff of Permian limestone; Permian limestone is well seen west of Doncaster, and at Mansfield in Nottinghamshire where the stone was quarried for the Houses of Parliament. In Ayrshire and Fife-shire there are stumps of Permian volcanoes. The sandstones consist of iron-coated, well-rounded grains of sand, clearly indicative of wind action and sand-dunes. In many places foot-prints of large animals, sun-cracks and pitting by rain bear witness to exposed tracts of sand and mud. As in Triassic rocks fossils are rare, and the stunted form of shells obtained from the limestone on the Durham coast is evidence of unfavourable conditions: beds of gypsum and salt are associated with some of the sedimentary beds. The Permian limestones differ from many other rocks of a similar kind in their pale yellow colour, and in consisting not solely of carbonate of lime but also of a lime-magnesium carbonate; they are called magnesian lime-stone or dolomite and were doubtless deposited in a partially enclosed and not in an open sea, in an environment analogous to the Caspian Sea. These magnesian limestones are characteristic of the eastern region in distinction to the much greater develop-ment of red sandstones to the west of the Pennines. York Minster is one of many buildings made of magnesian limestone. The red sandstones of Penrith and the Durham coast limestone are proof of contrasted conditions of rock building on the two sides of the Pennine Chain. At many localities there is another kind of rock, largely made up of angular fragments of many sizes, known as a breccia to distinguish it from a conglomerate consisting of well-rounded pebbles. The Permian breccias are highly resistant to the weather and stand out as ridges, e.g. the

Clent Hills south-west of Birmingham and the Abberley Hills; they were probably spread by water as fans of detritus in an arid country, as were the breccias of the Triassic period.

In the Permian period nearly the whole of Scotland was land; an arm of the sea lay over north-east England as far as the line of the Pennine Hills, over which there may have been communication to an almost closed sea over Cumberland, Westmorland and Lancashire and part of north-east Ireland. There was a lake in the Midlands and probably another over part of the English Channel reaching to Devonshire and the edges of the Bristol Channel. Into these seas and lakes torrential streams carried relatively large angular and rounded stones from neighbouring hill slopes, and quantities of finer rock-debris. The climate was semi-arid and there is little doubt that some of the Permian beds consist of sand derived from dunes. Conglomerates pass laterally into sandstones and marls: this means that where the conglomerates or old pebble-beds occur the rocks at the edge of the basin must have been close at hand; the sands made of finer-grained material are sediment carried a longer distance from the source of origin. The breccias associated with the sandstones and marls are probably scree deposits from the flanks of hills swept by swollen torrents into a neighbouring body of water. From an examination of the conglomerate pebbles and their fossils it is sometimes possible to trace them to the parent source: many were derived from rocks not very much older than those of Permian age, but on the other hand the breccias are made of angular stones from much more ancient rocks which, it is probable, formed part of a range of mountains that stretched from west to east across southern England to the continent. Remnants remain in the Mendips and in South Wales, but for the most part the great transverse ridge lies buried beneath newer rocks where its presence has been proved by boring; it is from this buried ridge that the Kent coal is obtained.

CHAPTER XIII

Forest, Delta, and Sea

As we pass in descending order from the rocks of one period to those of the next, it is not always apparent why geologists treat them as two chapters of earth-history rather than one. The rocks through which we have just passed are spoken of as Permian and those described in this chapter are assigned to the Carboniferous period, a distinction implying a natural line of division, a passage from one set of events to another. Let us briefly consider some of the reasons for this distinction. Sedimentary beds included in the Permian period usually lie at rather a different angle to those on which they rest, a discordance (see Chapter v, p. 65) which, though slight, is evidence of an interval in time during which the earth's crust in Britain was subjected to folding and dislocation. The difference in position of the Permian and Carboniferous rocks in relation to one another is technically described as an unconformity, a lack of parallelism which is a sign of a break in continuity. There is no clearly defined difference in the life of the lowest and oldest stage of the Permian period and the uppermost stage of the Carboniferous, but facts from many sources point to radical changes in the physical features of western Europe between the two periods. We shall consider later what happened at the close of the Carboniferous period; all that need be said now is that it was an age of mountain-building, when some of the more striking scenic features of Britain had their origin. The grounds on which this statement rests will be appreciated when we have taken a general survey of that period of ancient history with which this chapter deals.

The name Carboniferous was chosen because it is from beds belonging to this period that nearly the whole of our coal is obtained; it is the coal-bearing age. In the course of the 80 million years or more covered by the period, the history of the British Isles as deciphered from the rocks and their fossils may be conveniently divided into three phases—forests, delta-conditions, and open sea. In some parts of Britain the upper-most Carboniferous rocks afford evidence of a coming change from a moist climate, favourable to luxuriant forest vegetation, to a more arid and less favourable climate: in other words, there are clear indications of an approaching change in physical conditions which became more widespread and more pro-nounced in the succeeding Permian period. By far the greater part of the rocks referred to the latest Carboniferous phase, known as the Coal Measures or the Coal Age, tell a remarkable story of forest, swamp, and lagoons revealed in a pile of sand-stones, shales, and beds of coal reaching a thickness of some few thousand feet. Below the Coal Measures there are 2000 ft. and more of coarser sandstones with intercalated beds of shale and occasionally thin layers of coal: these rocks are mainly sediments from large deltas and are the records of the second phase known as the Millstone Grit from the former use of the gritty rocks as millstones for grinding corn. The first stage of the Car-boniferous period is that spoken of as the Carboniferous Lime-stone series; the rocks are mainly limestones containing a vast collection of shells and other hard parts of animals that lived in a clear sea and a genial climate. In some localities the oldest beds at the base of the limestone are conglomerates marking the position of old shingle beaches near a coast-line. The three phases are most clearly shown by the rocks in England; in Scotland, as we shall see, the physical conditions were not quite the same.

Carboniferous rocks cover a substantial proportion of the

British Isles: the geological map shows that they occur over the greater part of Ireland; in Scotland they are well developed in the Lowland or Midland valley. The greatest extent of Carboniferous rocks exposed at the surface in England is from the Scottish Border through Northumberland, Cumberland, Westmorland, Durham, Lancashire, Yorkshire, Nottinghamshire and Derbyshire; smaller exposures, which play a conspicuous part in the landscape, are in North and South Wales and the Bristol Channel district. There are many other places where the occurrence of Carboniferous rocks, buried below newer formations, has been discovered by borings.

The uppermost series, or Coal Measures, consists of sands and mud that were deposited in fresh-water lagoons, deltas, and estuaries and, in close association with them, at many different levels, seams of coal. In some places there are beds of red sandstone, very similar to the red Permian and Triassic rocks, at the upper limit of the Coal Measures, which show that the stage was being set for the next act, with an arid background, of the great drama. During by far the greater part of the Coal Age—probably at least a few million years—a large area was either land or submerged under large sheets of fresh water and, occasionally, under an invading sea. The alternation of sandstones, shales, and seams of coal through several thousand feet of rocks is proof of long-continued movement of the earth's surface, a rhythmic process of subsidence alternating with comparative stability. A bed or seam of coal is made up very largely of the vegetable litter of forests derived from generations of trees and other plants which succeeded one another over a long period and were gradually converted by pressure and chemical change into coal. The accumulation of forest debris continued without interruption until a pronounced subsidence of the ground caused an influx of water which drowned the forest: over the inundated forest were spread layers of sandy and muddy sediment that subsequently became beds of sandstone and shale. Eventually the deposition of sediment from the turbid floods came to an end; swamps replaced lagoons and lakes, and, as the area became silted up, the altered conditions made possible another colonization of the surface by other trees and humbler plants: a new forest phase was thus inaugurated. This alternation of forest, water, and swamp recurred time after time as cycle followed cycle in the spasmodic sinking of the restless crust.

On a geological map the Coal Measures are seen to occupy separate and not continuous areas; they occur in coal-fields, or coal basins as they are called, in which the rocks do not lie in horizontal layers over long distances, but are inclined at varying angles. On the west side of the Pennine axis there are the coal-fields of North Wales, North Staffordshire and Lancashire, and in Cumberland there is another patch of Coal Measures at Whitehaven, where part of the coal is worked under the sea. To the east of the Pennines are the coal-fields of Northumberland and Durham and of Yorkshire, Nottinghamshire and Derbyshire. There are coal-bearing beds in southern Scotland, in the English Midlands especially around Birmingham and, in a separate basin, the large coal-field of South Wales. Not very long ago coal mines were sunk in Kent through the Chalk to the relic of a buried mountain range, which consists in part of Carboniferous rocks. Allusion has already been made to the difference in degree of tilting of the uppermost Carboniferous and the lowest Permian beds, and it is this unconformity that enabled geologists to assign a date to the crustal disturbance. This corrugation of the crust brought into being two mountain ranges, one lying transversely across the south of Ireland and South Wales, and, as we know from boring operations, continued into south-eastern England and farther east into Belgium and beyond. As the folded rocks slowly emerged from the depths in which they were formed, they became a prey to denudation and the mountains were planed down to their present state. It is interesting to note that the continuation of this transverse ridge below the Chalk in Kent was suspected long before it was proved: having observed folding in the older rocks of Cornwall and Devonshire and in the Mendips, geologists prophesied from these features noticed at the surface that bores sunk below the Chalk of the Downs would demonstrate an eastward extension. This prophecy was confirmed when coal was discovered below Shakespeare's Cliff near Dover.

The folding of the crust along the east and west axis was not the only result of the disturbance which marked the end of the

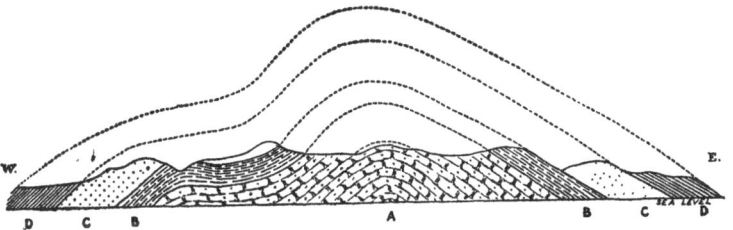

Fig. 8. Section across Derbyshire.

Carboniferous period: there was another folding which produced a second mountain range along a north and south axis caused by pressure acting upon the crust in an east–west direction. These foldings are comparable to the wrinkling of a table-cloth by the application of pressure. The Pennine Hills are the visible expression of the folding along a north–south axis. These two ranges—the Pennine Hills and the transverse ridge in the south at right angles to one another—provide an explanation of the occurrence of the Coal Measures in separate basins. A section across Derbyshire from west to east, based upon observations of the lie of the rocks at the surface, shows that the central core is a broad arch of Carboniferous Limestone (A) above which (B being a mixture of grits and limestone) the

14-2

Millstone Grit (*C*) and the Coal Measures (*D*) lie in regular order (Fig. 8). As the Pennine arch was being gradually uplifted, the upper beds were destroyed by the long-continued operation of rain and frost, and it was only in the less exposed troughs that the higher layers represented by the Coal Measures were left: hence their absence on the higher ground, and their preservation in basins where they are left as disconnected protected portions of rocks that were once continuous. Examples of folded and highly inclined Carboniferous rocks can be seen at many localities, on the faces of cliffs, in railway cuttings and elsewhere; one of the most impressive examples can be seen at Saundersfoot on the Pembrokeshire coast, where a steep pointed arch made of layers of sandstone of the Coal Measures rises on the edge of the beach: between the arch and the sea, where the waves have worn away the greater part of the rock, the remnants of the folded layers remain as low parallel ridges, those on one side sloping in one direction, those on the other side sloping in the opposite direction.

Let us now see what can be learnt from seams of coal which give to the uppermost division of the Carboniferous period its great economic value. In some places a bed of coal is visible at the surface on the face of sloping ground and elsewhere, and can be worked by tunnelling along its course, but more frequently coal is mined underground, and shafts are sunk through the accompanying sandstones and shales. In nearly every instance a seam of coal rests on a foundation of siliceous rock or clay, devoid of layering, and containing stumps of trees from which branches bearing numerous slender rootlets spread radially and almost horizontally. This foundation rock is called gannister when hard, underclay or seat earth when soft, and is the original surface-soil, altered by chemical action and weathering, the soil which was the floor of forests that were the source of coal. Occasionally a tree stem with its underground limbs embedded in the underclay passes vertically upwards through the actual coal, but usually there is a line of division between the coal and the old surface-soil. The stems and roots, being underground branches, are not preserved as wood, but generally as what are called casts, that is, they consist of sandstone or shale, which filled cavities left in the sediment or soil after the decay of the plant tissues, and show on their surface the markings on the outer part of the trees. A note-worthy fact is the very small amount of mineral matter—sand and mud—in coal, which consists of about 93 per cent of carbon and differs widely from ordinary sediment deposited from water. One can often detect on a piece of coal a more or less distinct pattern which recalls the surface-markings of a plant stem. Microscopical examination of fragments of the more sooty and softer layers reveals the presence of pieces of vegetable tissue, and sections of the harder coal are seen to contain other traces of plant structure, particularly flattened spores, the small repro-ductive cells such as those produced by club-mosses (*Lycopo-dium*), ferns and other plants. The greater part of coal is a black, dull or lustrous substance without obvious structure, but pieces of wood and bark are easily detected when thin sections are examined. The carbon of which coal is made was originally a constituent with oxygen of the gas carbon dioxide that was present in small amounts in the atmosphere of the Coal Age as it is in the air to-day. Living plants extract carbon from the air, which passes into the green leaves through countless minute pores: the green colouring matter, chlorophyll, has the re-markable property of absorbing rays of light and thus provides the living cells with a source of energy by which the cells are able to dissociate the carbon from the oxygen of the gas carbon dioxide. The carbon enters into a new combination with the elements oxygen and hydrogen and thus carbohydrates, such as sugars and starch, are built up, and these substances are rich stores of energy: the initial factor in their formation was sunlight trapped by minute living particles of chlorophyll in the leaves. Coal is not made of carbohydrates but of carbon; and the raw material from which it was produced was a kind of peaty mass consisting of a heterogeneous accumulation of vegetable debris on the floor of a forest. This material was covered with mud or sand deposited from the water that overwhelmed and killed the forest plants. Chemical action operating over long periods effected many changes in the forest litter and, eventually, little was left save the carbon that was part of a gas in the atmosphere in which the trees of the coal lived about 250 million years ago. The burning of coal is accompanied by a release of energy that has lain dormant through the ages, energy that came from the sun.

There is another very fruitful source of information on the nature of the plants of the Coal Age, in addition to that furnished by the coal itself; in some localities, especially in the coal-fields of Lancashire and Yorkshire, seams of coal contain hard, more or less spherical, stones varying in size from an inch to a foot in diameter. The so-called Halifax Hard Bed is a seam from which many of these were obtained, and it was by microscopical examination of the contained plant fragments that Professor W. C. Williamson, of Manchester, was able to lay the foundations of palaeobotanical science in this country. These stones are mainly composed of carbonate of lime and are known as coal balls or calcareous nodules: when examined microscopically in thin transparent sections they are found to contain pieces of a great variety of plants, twigs, leaves, seeds, etc., almost perfectly preserved by petrifaction. Another method of examining coal balls is to break one of them and make a smooth surface by grinding; this is then treated with hydrochloric acid which dissolves the limy part of the ball and leaves uninjured the plant tissue that is left slightly projecting from the stone. Cellulose or gelatine is then spread over the smoothed face and when it sets it can be detached as a transparent film with the plant tissues embedded in it and mounted on a slide for microscopical examination. Coal balls are, no doubt, petrified patches of the peaty substance most of which ultimately became coal; only the stony pieces remained as samples of the original vegetable debris. In the course of conversion of the peaty mass into coal there was a very great reduction in bulk consequent on compression of the soft vegetable material, but the coal balls show us patches of the raw material as it was. It is significant that immediately above a seam of coal containing coal balls one often finds a bed of shale with fossil shells: water percolating through the shale when it was soft mud dissolved some of the lime from the shells and redeposited it in the peat, the plant tissues being replaced, particle by particle, by indestructible stone. An examination of sections of coal balls enables us to see rootlets petrified in the act of boring their way through pieces of stems and other fragments, showing that roots were able to wander through the peaty mass. The conclusion reached is that by far the greater number of coal seams are made of the remains of plants which grew in the districts where the coal occurs; the coal was formed *in situ* on the actual site of the forests, and the underclay below each seam is the old surface-soil containing stumps and roots of some of the forest trees. The forests were on low ground that was often swampy and, from time to time, they were covered by incursions of fresh water, more rarely by the sea, which preserved the tangled mass of vegetation by sealing it up with sediment. This process was repeated many times, as is shown by the numerous coal seams intercalated among sandstone and shale in every coal-field. It is easy to collect fossil plants from blocks of shale brought from the underground galleries and dumped in heaps near the collieries: the fossils are often preserved as thin films of a coaly substance on the surface of the shale and include many kinds of fronds, superficially like those of living ferns, and pieces of flattened stems with various surface-markings, seeds and other odds and ends. The black film is all that remains of the plant tissues which were slowly decomposed and are now represented by a thin layer of carbonaceous matter. Water flowing over the forest-covered ground would carry with it samples of the vegetation as rivers now transport leaves and broken stems which happen to fall into them. No better example could be given of the value of work done by enthusiastic and painstaking collectors than the contributions to knowledge of the botany of the Coal Measures made by David Davies, an amateur, who in his spare time made a systematic collection of over 200,000 specimens in the South Wales coal-field. Having carefully noted the locality at which each specimen was collected and the level in the Coal Measures, he was able to follow fluctuations in the plant population from one stage to another through the whole succession of seams in the district; his devotion was rewarded by results of considerable interest and importance.

At recurrent periods during the millions of years of the Coal Age, the history of which is recorded in the Coal Measures, a considerable part of Great Britain was overspread by luxuriant forests, wide stretches of low-lying, swampy ground with lakes and lagoons, and here and there ridges of hills. Then, as now, there were Caledonian Highlands and, across the centre of England, a ridge of mountains, remnants of which are represented by the rocks of Charnwood Forest: the land that is now Ireland, southern Scotland and most of England was either forest-clad or submerged under shallow water. When forests had possession of the land the landscape was probably very similar to scenes in the Norfolk Broads; rivers flowing sluggishly through a low, swampy region with occasional lakes and lagoons, the open sea not far away; trees and shrubs and smaller plants of the undergrowth spread as a carpet of many shades of green over a substratum of black, oozy peat many feet in depth. The Dismal Swamp of Virginia also helps us to visualize the stagnant waters and slow-flowing rivers of the Coal Age forests. A natural question is—in what respects were the forests of the Coal Age different from those of the present day? Remembering that the Coal Age takes us back at least 250 million years, it is not surprising to find in the forests many trees that are strange and differ widely from any in the world we know; and yet the strangeness is seen to diminish a little when the fossils are more closely examined and we find in their structure characters reminding us of living trees. The tallest trees probably grew to a height of 100 ft. and more and may be described, in very general terms, as huge, overgrown club-mosses differing considerably in size, in their more robust and woody stems, and in other features from the small *Lycopodium* of British hills and the delicate *Selaginella*, allied to our club-mosses, that are familiar in greenhouses. These two genera are not mosses but allies of ferns. Other trees were constructed on an architectural plan reminding us of greatly enlarged and woody horsetails (*Equisetum*), but in many respects sharply contrasted. There were many plants bearing fronds very like those of ferns in size and form; most of these were not the leaves of true ferns but of members of a wholly extinct group of plants which were more advanced than ferns and reproduced themselves like our cone-bearing and broad-leaved trees by seeds and not merely by spores. A question that is often asked is—what light do the plants from the Coal Measures throw on the great problem of evolution? There were many plants too remote in structure from any that still exist to find a place in a system of classification designed to fit the present vegetable kingdom. In the forests were many representatives of classes that have long since disappeared, they are examples of Nature's experiments lacking the equipment necessary for permanence; there were other plants more or less similar to living forms which illustrate remarkable persistence through the ages. The forests of the Coal Age, so far as is known, contained no flowering plants, no oaks, elms, beeches and other trees that play so prominent a part in the modern world; there were no herbaceous plants brightening the woodlands with flowers. In the Carboniferous period there were no mammals; so far as we know it was not until a later age that the progenitors of the class that is now the highest in the animal kingdom were evolved. The most highly developed animals were creatures that have long been extinct; they were salamanders, some 4 ft. long, beasts that walked over the swamps or floundered in lagoons and rivers, or swam like eels. Insects of many kinds were abundant, dragon-flies, spiders, cockroaches and others, all differing in varying degrees from living species. No birds made music in the forests; their day had not yet come.

MILLSTONE GRIT

The sandstones, shales and seams of coal grouped together as the Coal Measures pass downwards into other sandstones, coarser grits, beds of shale and a few layers of coal reaching a thickness of several hundred feet which are referred to the next stage of the Carboniferous period and spoken of as the Millstone Grit series. The rocks of this phase are exposed at the surface in southern Scotland and reach their greatest development in Lancashire and Yorkshire; they are not found in the Midlands, but beds of similar type occur still farther south. In Yorkshire, from Richmond to Leeds, and in certain districts in Derbyshire, many of the hills are built of typical Millstone Grit, interbedded with layers of much softer shale. Rocks of this age underlie much of the north of England moorland: prominent escarpments are a characteristic feature of the Pennine Chain; they are recognizable by their gently sloping surface that follows the inclination to the horizontal of the beds of grit; a characteristic feature is their abrupt termination in a steep cliff-like scarp rising above lower wooded ground underlain by beds of less resistant shale. A river flowing over a district where grit rests on shale carves a channel through the hard rock and, when it reaches the more easily eroded shale below, the river at once cuts laterally, as well as downwards. Good examples of Millstone Grit can be seen in the Plumpton rocks, near Harrogate, with Scots pines growing on their weather-beaten surface; the Brimham rocks, near Pateley Bridge; and the furrowed Devil's Arrows near Boroughbridge. Kirkstall Abbey, near Leeds, is one of very many buildings built of stone from local quarries in Millstone Grit. The hills near Lancaster and farther south are capped by sloping escarpments associated with beds of shale.

Fossils in the Millstone Grit, though fairly common at some localities, are not abundant, and indeed rare, on the whole: casts of plant stems occur occasionally in the sandy rocks.

If one looks at exposures of Millstone Grit, as, for example, the large blocks frequently standing in groups on the moorlands, one notices well-marked sets of roughly parallel lines disposed at different angles, which have been made conspicuous by weathering agents; these are examples of current- or false-bedding (see Chapter IV, p. 45). A good many years ago a geologist, who made a thorough examination of the material of which Millstone Grit consists, found that it is a mixture of large and small grains of sand with milky white rounded pieces of quartz, and pieces of felspar, the mineral that occurs as large pink or white crystals in granite. A significant fact is that the pieces of felspar are comparatively fresh: under moist conditions, felspar soon disintegrates into clay, and its preservation in a recognizable state is probably due to the deposition of the grit in a rather dry climate. Both felspar and quartz are constituents of granitic rocks: it was therefore reasonable to conclude that such rocks may well have been the source of the sandy sediment. Remembering that the Millstone Grit attains a thickness in some places of well over 2000 ft., it is difficult to visualize rock destruction by rain, frost, and wind, on a scale sufficient to account for the formation of piles of detritus as thick and as widely spread as the Carboniferous grits and sandstones. When the grit was being spread by rivers over Britain, there was a large continent to the north, embracing the Scottish Highlands and the mountains that are now part of Scandinavia, mountains in which granitic rocks were abundant. Broad and swiftly flowing rivers, cutting their way through the Highlands, carried in suspension coarse and fine sediment derived from the wearing away by attrition of the rocks of the valleys and adjacent hills. This material was transported to the lower reaches of the watercourses where the force of the river was still strong enough to hold in suspension small pebbles, together with coarse and fine sand; by degrees, the layers of sediment deposited over the floor of the estuary, as the momentum of the rivers slackened, steadily increased in quantity and spread fanwise farther and farther over the shallows. From time to time, as the level of the ground slightly altered, the same area of deposition received still finer sediment, light enough to be held longer in suspension, which settled as sheets of mud and clay and subsequently became beds of shale. Thus, in the growing delta, layer after layer was added to the pile of sediment carried from sources far to the north and north-west. It is for this reason that the rocks of the Millstone Grit series are spoken of as the delta phase of the Carboniferous period.

CARBONIFEROUS LIMESTONE

Below the Millstone Grit in Yorkshire and Lancashire and other parts of England we come to a succession of beds made up of thinner layers of sandstone and shale with intercalated

rocks, partly sandy, partly calcareous, that are impure limestones, and other layers, in which there is little or no admixture
of sand, containing shells of marine animals. These rocks register
a passage from deep to shallower water, a change from a clear
sea to estuarine or delta conditions: this means a rise of the
earth's surface, an oscillation in level from shallow water near
the edge of a continent to deeper water farther from the coast.
The deltas of the river had gradually replaced a sea; a change
in level converted a sea-floor, sometimes deep enough to be
beyond the reach of sediment, into a region of sandy flats and
muds. Earlier in time conditions were more stable; the greater
part of England, practically the whole of Ireland and much of
Wales were overwhelmed by sea, a sea shallowing towards the
north of England and southern Scotland and deepening over
Westmorland, Lancashire, and Yorkshire. We have now reached
the open-sea phase, known as the Carboniferous Limestone
series. Thick beds of this limestone form the central arch of the
Pennine range and part of the more southern ridge seen in the
Mendip Hills, in the Cheddar Gorge, and in the Avon Gorge;
it plays a prominent part in the scenery in many parts of
England, North Wales, and Ireland; the great blocks of Ingleborough, Pennyghent and Whernside in the West Riding of
Yorkshire, the white cliffs in Derbyshire, Miller's Dale, Dove-
dale and other valleys, the hills on the north-eastern border of
Morecambe Bay, at Arnside, Grange and other localities. The
same limestones occur in places almost all round the Lake
District; they are the rocks of Great Orme's Head in North
Wales. Long exposure of the rocks to the solvent action of
water has led to the formation of long and tortuous caves, such
as those in the Craven district of Yorkshire, the Cheddar Cliffs
and elsewhere, also to the production of deep underground
watercourses, the best example of which is Gaping Ghyll, south
of Ingleborough. Another feature of many limestone districts

is the intersecting parallel series of fissures made by the widening of the joint-planes, by solution of their walls and the dissection of the rocky surface into channelled plateaux, with hart's tongue ferns lining the sides of the clefts. In some localities the limestone is seen to rest on a base of old shingle beaches which are indicative of the positions of coast-lines on the borders of the Carboniferous sea. The position of the margin of the sea in which the Carboniferous Limestone was formed from shells, corals, and other calcareous material is shown, not only by the occurrence of conglomerates or shingle beaches at the base of the limestone, but by a gradual thinning of the beds as the old land is approached. As already stated, the Carboniferous sea was not continuous over the whole of England; there was a belt of land across the middle of the English region, and one piece of evidence of this is the gradual decrease in thickness as the limestone is traced towards the edge of the barrier: in Derbyshire, the rock reaches a thickness of about 1500 ft.; in Leicestershire, this is reduced to 850 ft. and nearer the old coast-line, it is only 25 ft. thick and is associated with pebble-beds.

Districts where the grey and white limestones are exposed are among the most attractive to collectors of fossils. In places where the limestone has been used for building walls, one can see on almost every stone shells of bivalves, pieces of coral and stems of sea-lilies. On polished slabs of the rock (marble) the structure of many kinds of coral is clearly displayed and recognizable in cross-section by the radially disposed plates partially filling the cup or cylindrical stems, and corresponding to the fleshy folds in the body of a sea-anemone. Sea-lilies are represented by broken segments of the long, jointed, calcareous stems, in some parts of the country known as St Cuthbert's beads. Among the commoner and larger fossil shells are those of an extinct bivalve, superficially resembling those of an oyster, which often occur in crowded masses that were shell-banks on the sea-floor. The material of which the limestone is made varies from place to place; coral reefs, shell-banks and, in some places, various seaweeds, which, like certain living species, had the power of extracting lime from sea water. In the description of Chalk, a rock comparable to the Carboniferous Limestone, reference was made to the work of an amateur geologist who, by collecting sea-urchins from different bands of the rock, was able to provide valuable evidence from which it was possible, both to gain an insight into the nature of evolutionary changes, and to obtain trustworthy data by which to date the various levels by the occurrence of certain species of urchins. Laborious collecting of corals and other animal fossils, through many hundred feet of Carboniferous Limestone, has contributed very valuable information of a similar kind. It is within the reach of a keen and observant collector to extract from fossil-bearing rocks, at a locality which can be frequently visited, facts of considerable scientific importance. To become a recognized authority on even a single genus is an ambition well worth consideration.

So far, little attention has been given in this chapter to the Carboniferous rocks of southern Scotland, which tell rather a different story and throw additional light on the physical environment. The threefold division of the period into Coal Measures, Millstone Grit and Carboniferous Limestone, typically illustrated in Lancashire, Yorkshire, and Derbyshire, does not hold in an equal degree for Northumberland and Scotland, where alternating beds of sandy and muddy sediment and seams of coal afford evidence of proximity to land. In southern Scotland there was much volcanic activity, as shown by many lava-flows, beds of ash and the stumps of volcanoes. North Berwick Law is one of several isolated hills which were once active volcanoes, pouring from their craters sluggish streams of lava and showers of ash: now, all that is left are the worn-down stumps formed chiefly of rocks that had crystallized within the central pipe of the cone. Traprain Law, the Bass Rock, Arthur's Seat on the outskirts of Edinburgh, are other examples of volcanic activity in the Carboniferous age. Stirling Castle stands on a mass of volcanic rock; the Kilpatrick Hills and the Campsie Fells, near Glasgow, are also volcanic in origin. On the northern shore of the Firth of Forth, at Burntisland, beds of volcanic ash are a storehouse of petrified botanical treasures; they are crowded with scraps of trees and smaller plants of a forest overwhelmed by showers of a heterogeneous collection of rock fragments thrown from a crater in an explosive phase of activity. Similar beds of ash occur at Dalmeny, near the Forth Bridge, and they, too, have furnished exceptionally well-preserved stems that were killed but not destroyed: the wonderful perfection of some of the more delicate tissues revealed by the microscope makes it difficult to realize that we are looking, not at a section cut from a living tree, but at a stem of a forest tree that was alive more than 250 million years ago. From volcanic beds of the same age at Laggan Bay, on the north-east coast of the island of Arran, many petrified stems of extinct trees have been described. We can picture the Scottish scene as closely resembling an Italian landscape; volcanoes with their flanks and base clothed with vegetation that had grown precariously during a quiescent interval and eventually, when the volcanic forces broke into activity, the greater part of the woodland was destroyed. The destruction was not complete; branches and other scraps of trees buried in the volcanic ash were turned into stone by the infiltration of water charged with a mineral preservative derived from the volcanic material. Volcanic activity was widespread in southern Scotland; in England, there was comparatively little, but it is interesting to find layers of volcanic rock, locally known as Toadstones, inter-bedded with Carboniferous Limestone in Derbyshire.

A brief description of a hill in the West Riding of Yorkshire may serve as an illustration of the interpretation of rocks as documents from which to reconstruct the past. Ingleborough, one of a group of three hills, is a familiar landmark; it is made of almost horizontal beds of Carboniferous rocks, the lower 600 ft. is limestone containing a varied assortment of marine fossils; above this basal plinth of uniform structure we come to several hundred feet of thinner and less pure beds of limestone associated with shales and sandstones upraised from a sea-floor that was slowly rising and so converted a clear sea into a sea where sediment from the land was intermixed with the limy ooze formed at a greater depth. Reaching the top of the hill, we find a flat platform of Millstone Grit (Fig. 9). From the base of Ingleborough to the summit we traverse rather more than 2000 ft. of Carboniferous rocks, pure limestone passing into

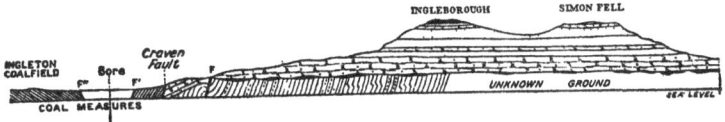

Fig. 9. Section across Ingleborough.

limestone containing sandy sediment, and layers of shale that were deposited as mud carried by rivers and currents; finally, at the upper limit, a thick covering of grit precisely the same as the much thicker beds of Millstone Grit in other districts of Yorkshire and elsewhere. The uppermost series—the Coal Measures—is lacking. A mile or two from the foot of Ingleborough is the country town of Ingleton with its small coal-field: below the coal-bearing beds, borings have demonstrated the presence of a considerable thickness of grits exactly like the flat platform on the top of Ingleborough. This shows that the highest and youngest rocks of Carboniferous age, though unrepresented on Ingleborough where the summit is formed of a bed of Millstone Grit, occur a few miles away on low ground more than 2000 ft. below the hill-top. How can this be explained? Imagine a plateau built up of limestone, grits and shales, all sediments from the bed of a Carboniferous sea of varying depth: the plateau was subjected to the action of subterranean forces from time to time set into operation by the disturbance of equilibrium within the earth's crust; the result of the pressure caused by the unleashing of these forces was to raise a considerable area of ground to a higher level, but the resistance of the uplifted rocks to the uplifting force was limited. Relief was eventually given by the formation of great cracks and fissures in the strained crust. If we push upwards from below a sheet of hard material, a slab of glass or slate for example, cracks will be caused when a strong enough force is applied and these will form lines of weakness along which there will tend to be slipping and a separation of the fractured mass, so that part of it, on one side of a crack, slides downwards and comes to rest at a level below that of the other part, which retains its original position. This is what happened at the end of the Carboniferous period: the crust of the earth was involved in widespread movement, there was what is called a crustal revolution; to this stupendous thrust the rocks responded by folding into arches and troughs, just as pressure applied to a tablecloth wrinkles its flat surface. In some parts of England, for example in Derbyshire, it has been shown that the Carboniferous Limestone and other rocks were forced upwards in consequence of lateral pressure into the Pennine range. In the Ingleborough district response to pressure was of a different kind; the rocks were not folded but were pushed upwards as great blocks of horizontally disposed layers riven by cracks and fissures many miles in length and thousands of feet deep. The rock thus raised formed a huge plateau; the fissures broke it into detached blocks: one of the largest of the fissures, known as the Craven Fault, has been traced for 80 miles running westwards past Settle, northwards past Kirkby Lonsdale and along the valley of the river Lune. The Craven Fault caused the Carboniferous beds to slip along the line of fracture, breaking the continuity of the rocks; on the Ingleton side the rocks slipped downwards, while those of Ingleborough and its companion hills remained at a much higher level: the difference in level on the two sides of the fault is over 5000 ft. The road from Giggleswick towards the village of Austwick runs between a limestone cliff overtopping the road on one side and much lower ground on the other side where rocks of the Millstone Grit are exposed: the road follows the line of the Craven Fault. An inspection of the rocks exposed on Chapel-le-Dale, Crummock Dale and other valleys in the Ingleborough district shows that the lowest horizontal layers of Carboniferous Limestone rest upon steeply folded rocks of much earlier date. This unconformity is evidence of a long period separating the two sets of beds: the older rocks were once part of a mountain chain that was raised by crustal movement from an early Palaeozoic sea; denudation wore down the ridge of high ground and, after millions of years, the surface of the land sank below sea-level and on it were subsequently built up the sediments which became the limestones, grits and shales of the Carboniferous period.

The marine fossils obtained from the Carboniferous Limestone are mainly bivalves, both brachiopods and lamellibranchs, corals of many kinds, sea-lilies, seaweeds with a calcareous covering, and several other animals of the sea. In addition to these extinct representatives of families and groups which have members in the oceans of the present age, there are also a few examples of a group that we have not so far met with, a group known as the trilobites, which is more fully described in a later chapter. The trilobites, as we shall see, were much more abundant and played a more important rôle in the older Palaeozoic seas; they had become greatly reduced in numbers and genera by the beginning of the Carboniferous period and it was in the seas of that age that they made their last appearance in Britain before finally disappearing during the Permian period.

CHAPTER XIV

Northern Lakes and a Southern Sea

Leaving the Carboniferous sea and the grey cliffs of the north and south-west of England that are the glorious monuments in stone of its long history and teeming life, we reach a chapter of history that is usually spoken of as Devonian, because it was in Devon that rocks of that age were described by geological pioneers. The Devonian period is in many respects a striking contrast to that which followed it: in its rocks are recorded two very different phases of history, two different sets of physical conditions over the region we now call the British Isles. In the sedimentary rocks of Devonshire, as in the Carboniferous Lime-stone, there are abundant corals and shells of other dwellers in salt water, a proof of the existence of an open sea seldom con-taminated by sediment from the land. These rocks give us a picture of a marine phase and a genial climate; but the great majority of the rocks from which the facts used in this chapter were obtained were deposited not in an open sea, but in inland seas and large fresh-water lakes. An important point to make at the outset is this: the documents of the Devonian period are preserved in two distinct series of rocks, one series typically represented in Devon, and essentially marine in origin; the other series, consisting of many thousand feet of sandstones, flagstones, shales and enormous piles of pebble-beds with a prevailing yellow and red colour associated with a considerable amount of igneous rock, has its greatest development in Scot-land and is known as the Old Red Sandstone. The Scottish stone-mason, Hugh Miller, a self-taught pioneer in the inter-pretation of the story of the rocks, was the first writer who, in his classic book *The Old Red Sandstone* (1841), made accessible to laymen the wonderful history of this period. In consequence of the much greater thickness and wider distribution of the rocks of this phase the term Old Red Sandstone is often applied to the whole Devonian period. It is, however, convenient occasionally to use the title Devonian in a restricted sense, that is, for the marine limestones and other sedimentary beds in the southern region where the land was covered with sea. This twofold character of the rocks of the Devonian period—using the name in the wider sense—is a characteristic feature of the British Isles; over the greater part of the European continent rocks of Devonian age are marine. In the British area it is only in south-west England, south of the Bristol Channel, that the marine type is found. The whole period covers at least 40 million years. Old Red Sandstone rocks, in contrast to many of the marine beds in Devonshire, are often barren and offer small rewards to the collector, but, for that reason, localities where animal and plant remains are known to occur are especially important and well worth attention; moreover, it is sometimes possible to discover fossils in places where they have never before been found. Throughout the greater part of the millions of years assigned to the Old Red Sandstone phase conditions remained fairly uniform; the rocks deposited include layers of sandstone and shale, heaps of pebble-beds, masses of material consisting of angular pieces of rock (breccias), and thick beds of lava and other rocks of igneous origin. During long periods sediments were slowly piled up on the floors of lakes; there were recurrent bursts of volcanic activity on land and under water. Comparison of fossil animals and plants collected at different levels through the thousands of feet of rock shows that, in the course of the inconceivably long period, the unfolding of life brought into being a succession of new forms illustrating progressive development, especially in the plant kingdom.

In the Shetland Islands, Old Red Sandstone rocks occur in the promontory south of Magnus Bay, on the west coast, and at Lerwick, on the east side; Fair Island, between the Shetland and Orkney Islands, famous for its pull-overs with characteristic designs said to have been given to the islanders by sailors of wrecked Armada ships, is a fragment of what was once a continuous surface of yellow and red beds; islands of the Orkney archipelago are built of Old Red Sandstone. At the southwestern corner of the Orkneys a conspicuous landmark is the Old Man of Hoy, a tall pillar of flaggy beds 450 ft. high resting on a plinth of igneous rock, detached from the main mass by Atlantic breakers (Plate VIII). Rocks of the same age cover part of the Caithness plain, to the east of the much older rocks of Sutherland, where they build the cliffs at Dunnet Head at the western end of the Pentland Firth, the impressive cliffs at Duncansby Head at the north-east corner of Scotland, past Wick to the Moray Firth and in patches to the east. On the south shore of the Moray Firth and between Inverness and Nairn pebble-beds show the position of an old beach at the foot of cliffs composed of rocks of much greater antiquity. Old Red Sandstone rocks are exposed over a diagonal tract of country along the northern boundary of the undulating lowlands of the Midland Valley, from Stonehaven on the coast about 20 miles south of Aberdeen to the Firth of Clyde and the island of Arran; this boundary is the line of the Great Highland Fault, a fracture which caused the Old Red Sandstone to slip downwards and brought it into juxtaposition with the more ancient border rocks of the Highlands. Most of the Midland Valley is occupied by Old Red Sandstone and Carboniferous rocks; on the southern border it is separated from the Southern Uplands by another fracture stretching from Dunbar to Girvan. Formerly the red rocks were, no doubt, spread over the land to the north and were mainly destroyed by denudation. The Ochil and Sidlaw Hills,

PLATE VIII

Photograph by York & Son

Old Man of Hoy

in the more northerly part of the Midland Valley, are prominent features due to the hardness of the volcanic rocks of which they are largely made. Similar volcanic rocks are associated with sedimentary beds in Glencoe and there are others between Loch Etive and Oban. The most southerly development in Scotland of Old Red Sandstone sediments and associated volcanic rocks is from Dunbar in the north to the Cheviot Hills over the border. In England rocks of this period occur only in patches until we reach Herefordshire, and the neighbouring counties on the south and west, where they underlie districts of rich cultivation, and, as at the Brecknock Beacon, form bold escarpments. Old Red Sandstone rocks do not occur in North Wales, but there are exposures in Anglesey, and in Shropshire and Radnorshire: in all probability they were formerly spread over most of the Welsh mountains. Old Red Sandstone rocks occur in a district south of Omagh in northern Ireland; the other groups are scattered over the large central area, by far the greater part of which is formed of Carboniferous Limestone. The largest development in Ireland of rocks of the Old Red Sandstone phase is a belt from Dingle Bay to Waterford, and it is from these rocks in southern Ireland that the best examples in the British Isles of Upper Devonian plants have been obtained, plants of a much higher type than those from the older rocks in Scotland. The northern shore of the Bristol Channel is the southern boundary of rocks of the Old Red Sandstone phase. For various reasons, partly differences in the animal and plant fossils, partly on physical grounds—the relation of the series of beds to one another—the 20,000 or 30,000 ft. of sandstones, shales, conglomerates, and other rocks, have been referred by geologists to separate subdivisions, Upper, Middle and Lower. For our present purpose it is unnecessary to draw a distinction between these stages, each of which represents many million years. The elevation of the Silurian sea-floor before the beginning of the

Devonian period converted part of the Scottish region into land with lakes and inland seas; one large lake extended from northern Ireland across the central Lowlands of Scotland, in which about 20,000 ft. of Old Red Sandstone sediments were deposited: over the bed of the lake were spread thick layers of lava and beds of volcanic ash, and summits of volcanoes rose above the water-level. This great lake covered a large Scottish area, and was bounded on the north by cliffs of a continent which included the North-West Highlands, and the highlands of Scandinavia. Land lay over the northern half of Wales and northern England, where there was a lake which reached into Scotland.

The enormous piles of Old Red Sandstone sediments are an impressive record of a stage in the history of the British Isles when the land was under more or less desert conditions; thick masses of water-worn pebbles derived from the old Highland rocks are evidence of occasional rain storms, of swollen rivers transporting heavy loads of large stones; masses of more angular stones suggest screes and fans of frost-riven rock fragments such as now are characteristic features in desert lands. Other evidences of an arid climate are the relatively small number of fossils, the well-rounded sand-grains like those of sand-dunes and the sand in desert countries, sun-cracks on sandstone that was once a sandy flat, and pits made by the impact of rain in recurrent storms.

The rocks of the marine or Devonian phase are confined to England, and are exposed in Devonshire from Lynton, where they are well shown in the Valley of the Rocks, to Barnstaple and Exmoor, and in South Devon eastwards to Plymouth and Torquay; they are seen at Berry Head near Torquay, at Dartmouth and at Totnes, and the valley of the Dart; at Anstey's Cove the rocks are folded and fractured and seem to be inextricably confused. Devonian rocks are widespread in Cornwall. The black jagged headlands of Morte Bay are highly inclined beds of sediment uplifted from the bed of the Devonian sea.

Hugh Miller made many valuable contributions to a know-ledge of the life in the Old Red Sandstone lakes, and of the vegetation that left scattered samples in the sediment derived from the land bordering the water. Since his day, prolonged and patient search and the application of improved methods of extracting from unpromising scraps of trees and shrubs the maximum amount of information have thrown much addi-tional light on the nature of the vegetation. Not very long ago a most important and welcome discovery was made at Rhynie in Aberdeenshire, of a bed of hard flinty rock full of beautiful petrified pieces of plants. The rock was probably formed by the deposition of mineral matter dissolved in the water of hot springs during a phase following volcanic eruptions; it may be described as a bed of peat converted into stone, containing plant remains and bodies of the oldest known insects. From these petrified scraps it has been possible to reconstruct some of the peat-forming plants which are among the oldest, though not quite the oldest, examples of land vegetation. Most of those preserved in the Rhynie Chert were not more than 6 or 8 in. high; the stem was slender and sparsely branched, some were leafless and rootless, obtaining water from the swampy ground by means of delicate hairs on the underground parts of the stem; others were clothed with small and crowded leaves, very similar to those of modern club-mosses (*Lycopodium*). The reproductive apparatus was very simple, the swollen ends of thin branches containing numerous small spores, almost identical with the spores of club-mosses and ferns. Though superficially comparable to some existing plants, they were peculiar in certain important features from which we can form some idea of early stages in the evolution of plants fitted for life on land. Many other plant remains have been collected from Old Red Sandstone beds in different parts of Scotland and the Orkney Islands, and from rocks occupying different positions in the order of succession of the rocks of the period. The plants from the lower subdivision are definitely simpler than those from the upper part of the Old Red Sandstone. For a fuller account of Devonian vegetation as a whole, reference must be made to other sources. So far as we know, there were no tall trees, certainly not in the earlier vegetation. A few of the simplest and lowest forms from the Rhynie Chert, which was formed in the middle stage of the period, agree closely with plants of the present age, but the great majority differ too widely from living members of the plant kingdom to be associated with them in the same group. It is, however, noteworthy that these ancient plants were constructed of cells precisely the same as the cells which make up the bodies of plants that exist to-day: then as now we feel certain that the living machinery drew its energy from sunlight, and it is reasonable to assume that the living protoplasm contained minute green bodies which absorbed rays of light. Many of the plants were partially submerged, and in some respects resembled seaweeds, while others were better adapted to life on land. It is possible—though proof is lacking—that the first plants to establish themselves on land came from older types which lived in the sea, and this possibility receives some support from the structure of certain of the older Devonian forms. The largest specimens of plants of the Devonian age in the British Isles were discovered in rocks of the upper series in southern Ireland: these include stems and cones of fairly tall trees that were ancestors of later types which lived in the forests of the Coal Age; large fern-like fronds, a few feet in length, bearing wedge-shaped or oval leaflets and capsules containing spores. These were in all probability the foliage of plants that were not true ferns, but early representatives of a class which later occupied a prominent place in Palaeozoic floras and persisted into the earlier stages of the Mesozoic era, the extinct class of Pteridosperms, that is fern-like plants which had advanced beyond the ferns and reproduced themselves by seeds.

Animal life in the Devonian period was of two distinct kinds: animals that lived in the open sea left shells and other hard parts in the limestones and other rocks of southern England; animals that lived in the large fresh-water lakes of the Old Red Sandstone phase introduce us to many strange creatures of a type not previously met with in our descent through the geological sequence. The marine fossils of the Devonian phase are, many of them, very similar to those characteristic of the Carboniferous Limestone: numerous corals, sea-lilies, some starfishes, many molluscs and other bivalves. Among the more interesting fossils from Old Red Sandstone rocks in Scotland and Herefordshire are primitive kinds of fish and animals, remote from any in the modern world and whose position in the animal kingdom is by no means certain. The fishes differed from ordinary kinds with which we are familiar, in having a cartilaginous rod along the back instead of the usual bony rod made of vertebrae; they were protected by a tough skin bearing tubercles and hard plates that formed an armour over the body. Some of these have their nearest relatives in the mud-fishes, such as *Ceratodus* of Australia and other rare survivals in Africa and South America, called also lung-fishes because they have not only gills but air-bladders that act as lungs enabling them to breathe when they are encased in mud during times of drought. They were forerunners of our common fishes and were descendants of still older Silurian species; some reached a length of several feet. A remarkable example of these extinct creatures was named *Pterichthys* because it had wing-like paddles attached to each front corner of the body-shield; it was discovered by Hugh Miller when he was working as a stone-mason in a quarry in Scotland. Another genus, *Palaeospondylus*, discovered in the flagstones of Caithness, is one of the oldest animals with any trace of vertebrae; it agrees in some of its characters with existing lampreys, which are a very low type of fish. The second group of Old Red Sandstone aquatic animals, often referred to as extinct crustacea, is illustrated by *Eurypterus*, *Pterygotus*, and other genera; they reached a length of 9 ft.; the body was covered with a chitinous membrane and they had large compound eyes. The existing king crabs (*Limulus*), familiar in aquaria, bear some resemblance to the larger Old Red Sandstone Eurypterids.

Devonian rocks in many places lie on highly tilted and folded beds of a much older period: this strongly marked discordance between two adjacent series, one almost horizontal and the other series inclined at high angles, helps us to reconstruct events preceding the initial stages of the Devonian period. At the end of the Silurian period, described in the next chapter, the earth's crust over a wide region of the Northern Hemisphere was subjected to far-reaching and internal strains and stresses which caused elevation and folding of the rocks into great arched and troughed ribs of high ground. This crustal disturbance, usually spoken of as the Caledonian revolution, may be described as the birth-pangs of mountain chains that, as the ground was heaved upwards and laterally compressed, rose to heights greater than the much more modern Alps of Switzerland. The building of these mountain chains reached its maximum before the dawn of the Devonian period; it was one of several cycles of crustal revolutions which were potent factors in changing the face of the earth. Not only were thousands of feet of sedimentary rock forced upwards from enormous troughs filled with water and containing sheets of sand, clay, and chalky ooze—the stuff of which rocks are made—that had been accumulating on the sea-floor through millions of years: there were other manifestations of pent-up energy that was released when the rocks, under terrific pressure, were no longer able to bear the strain. We can picture in faint outlines the unequal contest between the rocks and the irresistible forces surging as an earth-storm through the crust, agitating it as oceans are lashed into rollers as the wind furies reach their maximum effort. At length, the end came; the rocks broke under the strain, riven by long and deep fissures along which lateral and vertical movement became possible. The faulting and displacement seen in the Scottish Highlands are permanent expressions of the revolution. Another accompaniment of the revolution was the outpouring of lava over large areas: in addition to the molten material from sub-terranean sources which reached the surface, other liquid and semi-liquid matter was impelled upwards from deep-seated reservoirs into the overlying rocks and, still far below the surface, cooled into a crystalline mosaic, gradually and under the pressure of the covering it had failed to penetrate. The granite at Shap, in Westmorland, is an example of a molten mass that became a coarsely crystalline rock under precisely these conditions and during the same Caledonian revolution.

So far, only a few of the effects of this pre-Devonian period of mountain-building have been mentioned. In the North-West Highlands of Scotland, the narrow strip along the coast from a short distance east of Cape Wrath to Loch Inver, the original order of the rocks has in places been reversed by internal folding, over-folding, and dislocation. Rocks were not only bent into regular arches and troughs but, as the compressing forces increased, arches were squeezed until their symmetrical limbs were pushed forward and downwards so that one side of the arch was depressed and overtopped the shortened limb in front; folds were overfolded and became recumbent folds. Under this gigantic strain rocks were shattered and fractured, some of the fractures being formed along a line inclined at a low angle to the horizontal; when the crust at last yielded to this irre-sistible force great blocks of it were thrust bodily along the fractured surface, and ultimately came to rest on the top of the rocks that had once been above them, and thus the order of age was reversed. These disruptive forces may be said to have laid the foundations of the structural features of the Scottish Highlands, and of some other parts of the British Isles. The upheaval of mountain ranges, accompanied by volcanic outbursts, and the uplifting of molten material from deeply seated reservoirs, brought into being a new land built of rocks older than those of the Devonian period. Of this ancient land little is left save the battered and worn-down rocks and mountains: it was upon these inconceivably old ribs of the earth's crust that the ceaseless sculpturing of Nature's tools impressed the Highland scene and the landscape, which as we see it with a vision sharpened by understanding, awakens our sense of awe, and brings us nearer to the infinite.

Another consequence of storms that swept through the crust of the earth was the alteration of sedimentary and other rocks by intense heat generated in the process of folding and compression. Rocks that were originally soft mud or clay, and sheets of volcanic ash spread over the ocean-bed, were baked and pressed and transformed into hard slates: the well-known property of many roofing slates to split almost indefinitely along parallel planes, known as cleavage-planes, is caused by the orderly arrangement of the particles, which is an expression of the pressure that rendered the rocks semi-plastic, and thus led to the particles placing themselves at right angles to the compelling force. This pre-Devonian crustal disturbance was by no means confined to the British, or even to the European, region; its effects are recorded in the rocks of North America, northern Greenland, and the Far East. The Caledonian revolution was not only an important early stage in the building of the framework of the British Isles: it was also a major factor responsible for the change of ordinary sediments into rocks of great economic value; sandstones, shales, and limestones were converted by heat and pressure into rocks resembling in their crystalline aspect those formed by igneous action; sandstones became hard quartzites, shales became slates, and limestones were metamorphosed into crystalline marble. It was only after many years of observation in the field, and prolonged and occasionally bitter controversy, that geologists were able to unravel some of the secrets of the Highlands, to see in imagination the strength and the results of the Caledonian revolution, and to appreciate the fact that under prolonged strains and stresses even the hardest and toughest rock is little more than clay in the hands of the potter.

CHAPTER XV

The Older Palaeozoic Seas

In the Devonian period, as we have seen, there were two contrasted scenes, two sets of very different geographical conditions, over the western edge of Europe: the region that is now the southern margin of England was submerged under an open sea warm enough for reef-forming animals to build banks of coral. Over a much more extensive area, from the Shetland Isles through Scotland and far into England, large fresh-water lakes received enormous quantities of coarse and fine sandy sediment carried by flood-water and more gently flowing rivers from a semi-arid continent. The next stage in our descent brings us to some thousands of feet of various kinds of sedimentary rocks, all of which were lifted up during widespread movements of the earth's crust from the floor of an older sea. More than a century ago Sir Roderick Murchison gave the name Silurian to the period immediately preceding the Devonian. The Silures were a British tribe which fought gallantly for liberty against the Roman invaders; they lived in the southern part of Wales and the Welsh Borderland. Devonian and Silurian rocks, distinguished from one another by their fossils, occur together in some places and their juxtaposition might be regarded as evidence of an almost uninterrupted transition from one period to the other: such close association of rocks belonging to two periods is, however, often misleading. Between the latter part of the Silurian and the early stage of the Devonian period there was a long interlude in which the orderly process of rock-formation was interrupted by one of the greatest cycles of mountain-building in the earth's history. The Caledonian revolution, described in Chapter XIV, swept as a devastating storm over many regions, crumpling, overturning, and fracturing the rocks and spreading disorder through thousands of feet of the foundation-stones. With our narrowly circumscribed outlook we are apt to think of this crustal upheaval, reconstructed through an intensive study of the rocks, as a sudden catastrophic event: it was doubtless a gradual process which, viewed in retrospect, gives the impression of an awe-inspiring and rapidly enacted cataclysm.

In this chapter are included brief accounts of rocks of three geological periods, which may be spoken of as a set of documents enabling us to visualize the course of history in the earlier half of the Palaeozoic era: the periods in descending order of age are—Silurian, Ordovician, and Cambrian; they embrace approximately 200 million years. Despite the large slice of geological time, it is convenient and simpler to treat them for present purposes as three similar episodes in one protracted act of a great drama. The rocks tell a similar story, the spread of open seas, inhabited by an abundant and varying host of relatively simple marine creatures, over a large part of the British Isles, sometimes a contracted sea, at other times an expanding ocean. The rocks of the three periods are mainly sediments, often much altered by heat and pressure, with intermixed lavas and beds of volcanic ash, in part poured out and showered over the surface of the land, for the most part on the sea-floor. In the Silurian and Cambrian seas there was little volcanic activity: on the other hand, the Ordovician period was an age of fire; many thousand feet of igneous material were furnished by volcanic centres of eruption.

The name Cambrian was given long ago by Professor Adam Sedgwick of Cambridge to the oldest rocks of the Palaeozoic era; the title was chosen because he first studied them in Wales (Cambria). In 1879 Professor Lapworth of Birmingham proposed the name Ordovician for a series of rocks, sedimentary and volcanic, intermediate in age between some previously classed respectively as Silurian and Cambrian: this change in classification was made because more thorough examination had shown that the division into two periods, Silurian and Cambrian, was inadequate, and did not give a true picture of the course of geological history (see Chapter v). The Ordovices were the last British tribe in Wales and the Borderland to yield to the Roman legions. It must not be assumed that the shorter space devoted to the sketch of these three periods is a measure of the difficulty of the interpretation of the records; as a matter of fact the complicated history of the events they cover has in recent years taxed the ingenuity of many geologists. My purpose is not to discuss the many subdivisions of the three periods, and their correlation in different regions, but rather to give a general account based on well-established facts, omitting technical details. Silurian, Ordovician, and Cambrian rocks are constituents of the oldest visible land in Great Britain. It was from this ancient land in Wales and Shropshire that some of the material was derived which went to the building of the newer rocks over most of England. Let us first take a cursory survey of the three periods before describing the rocks in the districts where they are best displayed. When we pass from the lowest and oldest Devonian rocks we come to many thousands of feet of conglomerate, sandstone, shale, and limestone, all of which were formed as layers of sediment on the floor of the Silurian sea. The occurrence of conglomerates, which are old shingle beaches, helps us to trace the position of the shore line: the Breidden Hills, the Longmynds, and the Malvern Hills, are remnants of a peninsula lying between a northern and a southern sea. It has been estimated that the Silurian period lasted nearly 30 million years; large areas in southern Scotland, Wales and England were under water; the Highlands of Scotland and

Scandinavia were united as parts of a northern continent. During some stages of the Silurian age a transverse extension of the main continental region lay across England from Yorkshire to South Wales, but this was partially submerged under a shallow sea as the ocean-bed in the English region rose to a higher level. During this period there was hardly any volcanic activity in the British region, only occasional outpouring of lava over the sea-floor in certain districts. It was essentially an age of almost uninterrupted piling up of sediment on the bed of a sea which rose and fell with oscillations of the crust. The geographical distribution of the rocks is evidence of the wide extent of the sea: it has been possible to recognize and correlate the various kinds of sediment, sandstones, shales, and limestones by the fossils they contain, all of which are marine. It is true that a few very imperfect fossil plants have been discovered, but they tell us nothing of the nature of the contemporary vegetation of the land. A short account is given on a later page of some exceptionally interesting Silurian land plants discovered a few years ago in Australia.

Penetrating farther down we meet with a succession of thousands of feet of Ordovician rocks, some of which in structure and origin are very much the same as the sedimentary beds of the Silurian age, and, like them, they were formed on the floor of a sea; it was a sea inhabited by creatures similar to those of the younger period, though not identical. The chief difference between the Ordovician and the Silurian period is the enormous mass of lava and volcanic ash intercalated among the Ordovician sediments. As we shall see, Silurian and Ordovician rocks compose practically the whole of the English Lake District. The Ordovician period lasted more than twice as long as the Silurian.

We finally reach the still older and much longer Cambrian period: here, too, the rocks are marine sediments, with shingle beaches resting on the flanks of the most ancient land in the British Isles from which the pebbles were made by waves beating on the Cambrian cliffs. The sandstones, shales and other sedimentary beds, though containing fossils peculiar to them, are similar types of rock to those of the Ordovician and Silurian periods. As in the Silurian, there was little volcanic activity. Throughout perhaps 100 million years of Cambrian time, large areas remained under water; there was sea over northern and western Scotland, over North and South Wales, parts of England, and reaching to Scandinavia on one side and North America on the other. The simplest plan is to take a rather fuller geographical survey of the rocks of the three periods, treating them to some extent in chronological order.

THE LAKE DISTRICT

The English Lake District, an approximately circular area 35 miles in diameter in the counties of Cumberland and Westmorland, has a special charm for lovers of the more kindly type of hilly country, and even a superficial knowledge of its geological history can hardly fail to supplement a purely aesthetic enjoyment, by providing a broader appreciation of its significance based on the results of geological interpretation of the rocks. The smoothed contours of many of the hills, broad U-shaped valleys and hanging valleys, many of the deep basins filled with lakes and tarns, mounds and ridges of boulder clay and other superficial deposits, are some of the physical features, almost the latest finger-prints of Nature that are legacies from the Ice Age. The major features, the contrast of hills and valleys, bold escarpments and sharp-edged precipices, with fans of screes at their feet, have been carved out of heterogeneous rock-masses by the ceaseless and, to us, almost imperceptible operation of denudation and erosion. The Lake District proper is built of thousands of feet of Silurian and Ordovician sedimentary and igneous rocks: the southern half consists of Silurian rocks, and the northern half of Ordovician rocks, most of them volcanic in origin. A girdle of Carboniferous Limestone, interrupted here and there, surrounds the central region; the light grey encircling wall is the remnant of once continuous sheets of calcareous sediment from the floor of the Carboniferous sea, which, in a far-off age, and for a long period, lay over the whole. On the south-western border, in the Furness district, in the cliffs at St Bees farther north, and in the valley of the Eden on the east, there are patches of red rocks of Permian and Triassic age, which are the geologically youngest components of the peripheral rim. All these encircling rocks are the worn-down and basal portions of thick layers of sedimentary material, ranging from Carboniferous to Triassic in age, that in earlier days were spread as a thick mantle of stone over the then completely hidden Silurian and Ordovician formations.

Let us try to follow the events which converted the enormously thick layers of rock originally lying horizontally, or nearly so, and in more or less regular order over a wide expanse of country into the Lake District as it is now. A relief map of Lakeland, such as one sees at some railway stations, shows that the lakes tend to lie in valleys radiating from the centre as though cut by rivers flowing down the slopes of a great mountain. This feature, together with other kinds of evidence, has led to the conclusion that the whole region was lifted up as a gigantic dome-shaped bulge pushed upwards, in the course of crustal disturbances by an impelling force below the surface. As elevation proceeded, the destructive action of erosion through countless years gradually stripped off the upper layers and, eventually, exposed the older Palaeozoic rocks which, in turn, fell a prey to the differential sculpturing by water and frost which acted unequally on the hard and soft material. When did this upheaval occur? It must have been subsequent to the deposition of the youngest rocks affected by the crustal disturbance, namely the red sediments of the Triassic period which form the uppermost courses of the denuded and partially destroyed encircling girdle. Probably what may be called the birth of Lakeland, the uplift of the block that was eventually fashioned into mountain, fell and dale, was one of the results of the stupendous crustal revolution in the earlier half of the Tertiary era, described in an earlier chapter as the Alpine storm.

We now turn to the rocks in the heart of the Lake District. Rocks of Silurian age make up the southern half, from Cartmel and Ulverston, in the south-west, to Tebay and Sedbergh, in the east, and as far north as the head of Windermere; Silurian beds are exposed in the Coniston district, at Ambleside, and over the whole of the lower and less rugged half of Lakeland. The rocks are shales, flagstones, beds of impure limestone and other sediments that were deposited on the floor of the sea. Some of the shales containing graptolites are rich in carbon and consequently black and dark grey in colour, and it may be that the carbon was derived from the decomposition of floating seaweeds to which graptolites had attached themselves. In the more calcareous beds, for example in the neighbourhood of Coniston and Ambleside, remains of other marine animals are abundant: the discovery of many species of starfish in one of the limestones is proof of the antiquity of these familiar creatures of modern seas.

The rocks of the Ordovician period include sedimentary beds similar to those of Silurian age, but they are subordinate in amount to many thousand feet of lava and volcanic ash. The northern half of the Lake District is made of Ordovician rocks: the oldest series—the Skiddaw Slates—consists of shallow-water mud and sand hardened and metamorphosed into slate. These beds are well developed in the Keswick district, Saddleback and

Skiddaw: the country on the shores of Bassenthwaite Lake is a typical example of Skiddaw Slate scenery; also, around Buttermere and Crummock Water. The rocks next above the Skiddaw Slates are referred to the Borrowdale series; they are mainly volcanic and reach a thickness of 20,000 ft. The Borrowdale rocks form a strip nearly 20 miles broad, running south-west and north-east, and including all the most spectacular parts of the Lake District, with Helvellyn, Scafell, the Langdale Pikes and most of the other high peaks and famous climbs. Near Grange, in Borrowdale, it is possible to see the junction between the Skiddaw Slates and the Borrowdale series, recognizable by the dark green colour of the volcanic ash in contrast to the blue and black Skiddaw Slates. From the west shore of Derwentwater one sees terraces of rock-layers on Falcon Crag which are due to beds of hard lava. Castle Head, near Keswick, is a prominent feature in the landscape caused by the presence of an igneous rock that was intruded into Ordovician beds. Veins of graphite and of copper, lead and zinc in the Skiddaw and Borrowdale districts occur in the Ordovician series, and possibly their formation may have been connected with the igneous activity. It is not always possible to say definitely whether lava-flows and volcanic ash were formed on a land-surface or the floor of a sea; the absence of fossils may mean that they were subaerial in origin, though some were, no doubt, submarine.

SOUTHERN UPLANDS OF SCOTLAND

The transverse belt of country in southern Scotland, beyond the English border, from the North Sea to the Irish Channel, is known as the Southern Uplands; it is bounded on the north by the fault which marks the southern limit of the Midland Valley or Central Lowlands, and on the south by the Solway Firth and the Cheviot Hills. Travelling to Scotland on the L.M.S. railway from Carlisle, one passes through typical Uplands scenery, a rolling sea of broad, rounded hills intersected by deep, narrow valleys. Most of the country is on Silurian and Ordovician rocks with some Old Red Sandstone. In the south-western district, in Galloway, including Criffel and Dalbeattie, large blocks of granite give a more rugged and grander character to the landscape. The Silurian and Ordovician rocks are mostly shale, sandstone and grit, sediments from the floor of a sea containing many species of graptolites: the rocks are steeply inclined—that is, they dip at a high angle—and their position is indicative of intense folding; they represent the basal wreck of an ancient mountain range which was one of the results of the Caledonian revolution at the end of the Silurian period. In the following description special attention is paid to the sedimentary beds of the Ordovician period because they serve to illustrate a brilliant piece of intensive geological exploration which provided a solution of a very difficult problem. Rocks of Silurian age are associated with the Ordovician beds in the Southern Uplands, and both were involved in the crustal folding responsible for the complicated structural features described below. Ordovician rocks are exposed on the west coast at Stranraer, Ballantrae, and Girvan: the beds of shale and the much thicker beds of grits are spoken of respectively, from the places where they are well developed, as the Moffat Shales and Girvan Grits. Comparison of fossil graptolites in these beds showed that both series belong to the same geological stage. The shales pass laterally into sandstones and grits, and this change in the nature of the sediments—shales becoming replaced by grits—is evidence of deeper water being replaced by shallower water. At Moffat the shales are 300 ft. thick; in the Girvan district they are represented, as shown by the comparison of fossils, by some thousands of feet of sand and grit. In Chapter v an account has already been given of the apparently inverted order of some of the beds in the Southern Uplands, and of how a study of the graptolites in them led to the conclusion that the rocks had been intensely folded into arches and troughs, and that subsequently the upper parts of the arches had been removed by erosion. The graptolites, so far as is known, died out in the Silurian period; they were very abundant, and ranged over the oceans of the world in the older Palaeozoic periods. It is recorded that an enthusiastic amateur geologist, Mrs Robert Gray, and her family, collected 30,000 specimens in the Girvan area alone.

The oldest rocks in the Cheviot Hills are Silurian flagstones, shales, and grits. In the West Riding of Yorkshire, highly inclined and folded beds of the same period underlie the almost horizontally layered Carboniferous limestones—at Horton-in-Ribblesdale and Austwick near Ingleborough the discordance between the rocks of these two periods is clearly shown. The Silurian rocks are part of an old worn-down, elevated tract, comparable to that in the Southern Uplands, which eventually sank beneath the sea, in which, long afterwards, the limestone was formed. Silurian rocks occur also in the Howgill Fells, and extend south to Kirkby Lonsdale.

WALES AND THE WELSH BORDERLAND

Large areas in North and Central Wales are underlain by Silurian rocks: at Plynlimmon they reach a thickness of 10,000 ft. and this affords a measure of the time taken in their formation as sheets of sediment on the sea-floor. Rocks of the same period are exposed at many localities in the Welsh Borderland, in Shropshire and farther south in Herefordshire, where beds of hard, shelly limestone stand out as conspicuous features in the landscape, for example, the limestone scarp at Wenlock Edge, 16 miles long, from which large collections of exceptionally well-preserved shells have been made. In the Ludlow district the abundance of corals shows the occurrence of reefs built up by generations of coral-forming animals. Pebble-beds at the

foot of the Malvern Hills, and on the eastern slopes of the
Mendips, mark the position of shingle beaches by the edge of
the Silurian sea. Other rocks of this age have been disclosed by
a boring at Ware in Hertfordshire, sunk through the younger
sedimentary beds at the surface, to a depth of 800 ft. into a
buried ridge of Palaeozoic rocks, which is the buried prolongation of Silurian beds exposed in Wales.

Ordovician rocks are exposed in many districts in North
Wales and there are some in the island of Anglesey: they are
seen at some localities to lie on the eroded surface of Cambrian
rocks, and this is evidence of discontinuity in geological history,
which implies that the upraised beds from the Cambrian sea
were exposed to denudation before the sedimentary rocks were
again submerged to form the floor of the Ordovician sea. As
in the Lake District, so also in Wales, there was much outpouring of lava, and scattering of volcanic ash from volcanic
islands or fissures in the ocean-bed. The occurrence of marine
fossils in beds of lava is proof of their marine origin. Cader
Idris is in part made of sediments, and in part of parallel and
highly inclined layers of volcanic rock: its prominence as a bold
escarpment is the result of resistance of the hard material to
weathering agents. Snowdon is another block of Ordovician
rocks, mostly lava and ash which, in contrast to those of Cader
Idris, are folded into a trough. One might expect to find a
mountain constructed of arched layers of rock, and not of
rocks bent down into a trough: but it is by no means rare
to find prominent portions of the earth's crust composed of
beds sloping inwards and downwards, and not upwards as
limbs of arched folds. This is no doubt mainly due to the
fact that arched beds are stretched while troughs are compressed: the tops of the arches are more liable to fracture,
and thus made an easier prey to the wasting action of rain and
ice. Ordovician rocks occur also at Wrekin, at Caer Caradoc,

and west of the Longmynd Hills in Shropshire, in the Breidden Hills in Montgomery, in the Hereford district, at Fishguard, and other places. The volcanic rocks in the Prescelly Mountains in Pembrokeshire are the source from which the 'foreign' stones of Stonehenge were taken.

Rocks of the Cambrian period also play a conspicuous part in the scenery of Wales: they are almost entirely sedimentary, conglomerates that tell us of proximity to land, with sandstones and shales. The sandstones have often been converted into hard quartzite by the metamorphism of the originally loose sand-grains into a compact crystalline mass, and similarly the slates that were once layers of clay on the sea-floor acquired their present structure and tendency to cleave under the transforming influence of heat and pressure, caused by crustal disturbances. The so-called Harlech dome, the arched mass of rocks in that district of North Wales, is an upraised, bent block of grits and flagstones. The slate quarries of Bethesda, and other places, are impressive illustrations of the enormous thickness of muddy sediment built up layer by layer in the course of millions of years, on the bed of the Cambrian sea: in a later age and under the stress of irresistible forces, the sedimentary material became plastic or semi-plastic and the rock particles were able to adjust themselves to the pressure by taking positions at right angles to the direction from which it came. Cambrian rocks made of pebbles and coarse grit resting on the eroded surface of the still older pre-Cambrian rocks in the cliff near St David's in South Wales enable us to visualize the waves of the Cambrian sea beating against cliffs that were part of some of the oldest land in Britain.

NORTH-WEST HIGHLANDS OF SCOTLAND

The North-West Highlands, separated from the Grampian Highlands by the great Glen Fault, which runs from the Moray

Firth to Loch Linnhe, is a region almost entirely constructed of rocks belonging to the pre-Cambrian era described in the next chapter. There are, however, several kinds of Cambrian rocks intimately associated, and in places, inextricably intermixed with those of pre-Cambrian age, along the inner border of a narrow strip of country, from Durness near Cape Wrath, on the north-west coast to the Sound of Sleat in the south, rather more than 100 miles in length. This strip of the North-West Highlands is separated by a fault- or thrust-plane from the very much larger part of the Northern Highlands, where the rocks are of a different kind, and most of them of pre-Cambrian age. The coastal strip, including part of the counties of Sutherland and Ross, is a tract of disturbed ground where the original relationship of one series of rocks to another has been rendered almost unrecognizable by movement and fracturing of the crust on a titanic scale, and for many years geologists failed to discover the relative ages of the rocks in this confused and fractured region. The lowest members of the Cambrian formation in this area are pebble-beds, formed by the wear and tear of the cliffs of pre-Cambrian land, which had been exposed to long-continued erosion by rain and frost, and, at a late stage, to denudation by an invading sea. Over this folded and eroded land the sea gradually encroached, and on its floor shingle beds, sand and deposits rich in carbonate of lime were laid down. The sandy sediment was ultimately transformed into white quartzite when the Cambrian beds were involved in the crustal revolution: this rock is in some places 200 ft. thick: it contains practically no fossils. The quartzite forms a cap of almost snow-white rock on Quinag and other mountains in the Loch Assynt district and elsewhere. The altered sandy rocks are followed at a higher level by beds of finer, muddy sediment: there are also beds of limestone of a considerable thickness in the Durness area in the extreme north. These Cambrian rocks are met with not only in the Assynt district, but at Loch Maree, and Ullapool farther south. The age of the beds is shown by shells of marine animals, and particularly by a trilobite called *Olenellus* which was discovered in 1888 in an early Cambrian bed in Western England, and later in the Durness Limestone in the North-West Highlands. The limestone had previously been classed as Silurian, and it was the discovery of the trilobite which proved its much greater antiquity.

There was no lack of life in Silurian, Ordovician, and Cambrian seas: each period had its characteristic set of species, but throughout the millions of years, as the water waxed and waned over the flooded land, the changes in the major features of animal life were comparatively slight. We have already seen that fishes of a primitive type abounded in the Devonian sea, over southern England, and some of them occur in the Silurian; so far as is known, that was the age when they made their first appearance in Britain. With this exception, the seas of the three periods contained no examples of vertebrate animals: none had yet come into being. A very few trilobites are recorded from the Permian, some from the Carboniferous, others lived in the Devonian sea, and still more in those of the Silurian, Ordovician, and Cambrian. This wholly extinct group reached its greatest development in the Silurian and Ordovician periods, and was represented by a great number of genera and species, varying in size from less than an inch in length, to about 18 in. The name 'trilobite' has reference to the three-lobed form of the body: the chitinous coat in most forms has a rather prominent central axis, like a more or less rounded, convex rib, with a flattened border on each side. The body is constructed of several overlapping transversely jointed segments, comparable with the joints in a lobster's shell: in exceptionally well-preserved specimens, slender legs and antennae have been found still attached to the outer case. Typical trilobites show a threefold division in a longitudinal direction into a head, a body, and a tail: specimens occasionally occur in which the whole body is rolled up like a wood-louse. Some had large compound eyes made up of as many as 15,000 facets. It is not easy to assign the trilobites to a precise position in the animal kingdom: they are probably nearer to scorpions and spiders than to any other living members of the class known as Arthropods, because of their possession of jointed feet. Reference has previously been made in Chapter v to another extinct group—the graptolites— which were confined to the Lower Palaeozoic seas: they, as well as the trilobites, were comparatively short-lived; new species were rapidly evolved, and each lasted only a short time; both groups are valuable as guides to geological horizons, serving as trustworthy index-fossils. Many of the shells and other hard parts of Lower Palaeozoic marine animals appear to be closely allied to genera that still exist: the bivalved brachio-pods were very much more numerous and more varied than in later periods. One of the most remarkable examples of a brachiopod genus which lived in Cambrian seas, and is still living in a few places in modern seas, is *Lingula*, probably the most impressive illustration of continuity and persistence in the animal kingdom.

The most interesting and significant fact that emerges from a survey of the life of the oldest of the three periods now under consideration is its variety; all the classes, except vertebrates which make up the population of present-day oceans, had already been evolved in the Cambrian period. This is con-vincing proof, or so it appears to those who accept evolution, that there must have been a long period antecedent to the Cam-brian, during which ancestral and simpler forms inhabited a much older sea. Unfortunately, the rocks of the pre-Cambrian era contain no records throwing any light on the precursors of the Cambrian animals. 'A time there was when life had never been', but that time was long before the dawn of the Cambrian age. The apparently sudden appearance in Cambrian seas of a great multitude of animals, most of them types relatively high in the scale of organization, means either that they are examples of sudden creation, or that they were the descendants of much more ancient types, of which no trace is left. A most important discovery of Cambrian fossils was made several years ago by a well-known American geologist, in sedimentary beds in the Rocky Mountains of Canada, who found in fine-grained mud, upraised from the sea-floor, a large number of beautifully preserved shells and skeletons of a great variety of genera, including some exceptionally perfect trilobites, and very many other extinct creatures, together with animals closely related to genera and species that still exist. Among the fossils were some that had no hard covering, but had left clear impressions of their soft bodies in the muddy sediment; several seaweeds of a simple type were also found. Would that we could find a similar deposit deep down in the pre-Cambrian rocks, and thus obtain a glimpse of the earlier stages in evolution which are as yet hidden from us.

What of plant life in the Silurian, Ordovician, and Cambrian periods? It is not to be expected that sediments from ocean-beds could tell us much of contemporary life on the continents on their margin. All we could expect are scraps of plants carried by rivers along with their load of sand and mud: plants which grew on river banks, or on the flat land of estuaries, within reach of transporting streams. A brief account of some of the oldest known Devonian plants, which grew on land or partially submerged, was given in Chapter XIV. A noteworthy fact is that certain Devonian plants have been found in widely separated parts of the world, in Scotland, Norway, North America, South Africa, China and elsewhere; this shows that they must have been in existence long enough for them to wander from one end of the world's surface to another, to have been spread by wind which carried their spores—the small reproductive germs—by slow stages over an enormous area. The few plant remains from Silurian rocks which have been discovered, are, nearly all of them, too obscure and doubtful to have any scientific value. A few years ago collections of fossil plants were made in Victoria, Australia, and fortunately some of them were found in close association with graptolites which were recognized as Silurian species, and therefore dated the rock. A detailed description of the fossils would of necessity be technical and tedious to non-botanical readers, but there are certain facts to which attention may be profitably directed. One of the Silurian plants is a species of a genus called *Zosterophyllum*; this closely resembles examples of the same genus previously found in early Devonian rocks in Scotland, and some other places; it was a comparatively small plant, not more than about 8 in. high; it had slender, occasionally branched, leafless stems, which were hardly strong enough to stand erect on land, and may have been supported by a partial covering of water in swamps. The upper end of the branches bore capsules containing spores of a kind suitable for dispersal in air rather than in water. *Zostero-phyllum* was no doubt a land plant, even though the lower part may have been in water: its name implies a resemblance to the living ribbon-like marine flowering plant *Zostera*, the grass-wrack, but there is no relationship between the two. This Devonian and Silurian plant is included by botanists, with some others of the oldest known types, in an extinct class confined to these two periods. The stem was provided with a simple central strand of what is called vascular tissue, that is, a group of very small tubes for the conduction of water absorbed from the ground to other parts of the plant body; this conducting strand is an attribute of plants growing on land. *Zosterophyllum* was not only without leaves, it was also without roots and absorbed the raw material in the soil by hairs. There were other Silurian plants: one of them (*Baragwanathia*) had much thicker stems which bore crowded long and slender, needle-like leaves, some of which bore at the base, close to the stem, a spore-capsule. Externally the plant must have looked like some existing club-mosses (*Lycopodium*), especially some tropical species, more robust than our British examples. This is clearly a plant of the land, and a type which may well be closely allied to the tall and more tree-like *Lepidodendra* of the coal forests. When we examine the oldest known plants that lived on land we have to admit that they indicate a stage in evolution which must have been far from the earliest phase of development of terrestrial vegetation. For an account of other Silurian plants reference should be made to original sources.

A few fossils from Silurian, Ordovician, and Cambrian rocks are undoubtedly remains of seaweeds: some of the most convincing are small, cylindrical and tubular stems, a few inches long, with many perforations marking the attachment of very slender branches. These are calcareous seaweeds, plants which extracted carbonate of lime from the water and covered their delicate bodies with a protective coat. The oldest known species of these seaweeds (Algae) from Ordovician and Cambrian rocks, agree closely enough with living forms of warm seas to be assigned to the same group, and are striking examples of persistent types which have maintained their general plan of construction, with only minor changes, from the Cambrian period until the present day.

CHAPTER XVI

The End of the Journey

We have now reached the last stage in our backward journey, the oldest volume, or, more correctly, series of volumes, the source from which geologists endeavour to piece together the fragmentary and often indecipherable records of the earth's youth. In some text-books the oldest era is subdivided into Archaean and pre-Cambrian, the lowest and most ancient rocks being grouped together as Archaean (Greek, *archaios* = ancient), the upper and younger as pre-Cambrian, that is, the rocks immediately below those of the Cambrian period. Many other names have been proposed for subdivisions of the two major groups, but with them we need not concern ourselves. It is preferable and simpler to use only the term pre-Cambrian in a comprehensive sense for all the rocks known to have been formed before the dawn of the Cambrian period. Many secrets hidden in the rocks of this era will never be discovered; some will doubtless be revealed to future investigators. We can, however, confidently assert that the pre-Cambrian era as a whole includes a pile of sedimentary and igneous rocks, several miles in thickness, and covers a period of time perhaps twice as great as the tale of years represented by all the other eras and periods put together. It used to be thought that some of the coarsely crystalline pre-Cambrian rocks referred to as Archaean were portions of the primaeval crust which solidified on a cooling molten mass, as the actual foundation-stones of the world, but several years ago it was found that the supposed original crust over a large area in Canada could not be so regarded because the rocks had been intruded as molten material into beds of still older sediments lying above them. The answer to the question 'Whereupon were the foundations thereof fastened?' has not yet been found. Sedimentary rocks, such as sandstones and shales, are made of the products of erosion of pre-existing rocks, and it is therefore clear that they could not be part of the original crust; they must have been produced from an older source. Neither the sedimentary rocks, nor those intruded into them, could be the earth's foundation-stones. The nature of the primaeval crust is therefore a matter of speculation. After prolonged loss of heat by radiation from the gaseous and molten earth, after its birth from the sun, a solid superficial envelope was gradually formed; portions of this may have sunk into the still molten matter below, and remelted. Eventually, greater stability was established, and a more permanent crustal skin covered the cooling surface. As the temperature continued to fall, steam condensed into water, which filled depressions on the barren foundation. The primaeval seas were fresh, and did not become salt until they received chemical substances from the breaking up of rocks, and decomposition of minerals. It would seem reasonable to assume that the first solid crust was an aggregate of crystals similar, no doubt, to rocks we know on the world's surface in the present age.

Rocks of pre-Cambrian age are exposed over broad regions in many parts of the world, and it is estimated that they cover about one-fifth of the whole land-surface. One of the largest blocks of coarsely crystalline rocks, mainly gneiss and granite, occupies a region of at least two million square miles from Labrador and the St Lawrence river, to Lake Superior, Lake Huron, the Hudson Bay district, and far into Arctic Canada: this so-called Laurentian Shield is the foundation of more than half of Canada. It is made up of an enormous thickness of sedimentary rocks, which in the Lake Superior region reach a depth of 50,000 ft.; also rocks of volcanic or igneous origin 30,000 ft. in thickness.

The largest pre-Cambrian shield in Europe, similar to that in Canada, embraces parts of Scandinavia, Finland, and the Kola peninsula; in Finland, as in Canada, granite rocks cover a very large area where hills, rounded and grooved by moving sheets of ice, and hundreds of lakes are the dominant features. Greenland is mainly built of the same kind of rock now covered by sheets of ice, except on the exposed coastal edge; at different geological periods the margins of the elevated plateau were overlain by thick layers of sediment deposited in estuaries and shallow seas during partial submergence under transgressing seas of later ages. Other pre-Cambrian shields are prominent features in Brazil, Africa, and other parts of the Southern Hemisphere. On a smaller scale, rocks of the same or approximately the same age are exposed in the central plateau of France, in the Black Forest, and in several other European countries. The rounded granite bosses and the fantastic piles of gigantic boulder-like blocks, cut from the parent rock by the action of the weather, are impressive features of a pre-Cambrian landscape in the Matopo Hills of South Africa, a region which has been part of the earth's surface almost since the beginning of geological history. It was there that Cecil Rhodes, empire builder and mystic, chose his burial place in a solitude, remote from the world of man, in a rock that may well have been part of lifeless Mother-earth. Pre-Cambrian rocks form part of the Alps, the Himalayas, and the other mountain ranges, intermixed with sedimentary and igneous material of later periods. Shields and knots of pre-Cambrian rocks enter into the composition of a vast area of the earth's surface, and are accessible to investigation: could we bore deep enough through regions where the surface is made of younger material, we should no doubt find a basal platform of coarsely crystalline granite and gneiss similar to the oldest known rocks of the exposed shields.

The name pre-Cambrian era must not be taken to imply a

continuous orderly sequence of events: the era included many periods which cannot be distinguished or classified with a precision such as is possible in the later volumes of geological history. The reason for this is that the oldest parts of the earth's crust have been repeatedly subjected to violent folding and fracturing, with the consequent overturning and disarrangement of the component layers: moreover, under the transforming influence of intense heat, accompanying the strains and stresses, many of the rocks must have been metamorphosed beyond recognition. It is therefore not surprising that a satisfactory chronological classification and correlation is hardly possible. In the course of the hundreds of millions of years of the whole era, there were several cycles of mountain-building, separated by periods of relative stability and quiescence: volcanic activity was widespread and recurrent; sedimentary material was deposited on an enormous scale, and from time to time, huge quantities of molten matter were forced upwards from the depths and slowly solidified under the pressure of overlying rocks, as coarsely crystalline domes and shields that were subsequently laid bare by erosion and denudation. In quiet intervals in the long succession of rock-construction and rock-destruction, beds of limestone grew in bulk on sea-floors. All these rocks were involved in repeated earth-storms, fractured and folded, and their structure altered by heat to such an extent that they lost those characteristics by which their manner of origin could be deduced. Pre-Cambrian rocks may be described as a heterogeneous complex, which by reason of their complexity and the fact that they are the sources from which alone it is possible to picture the earliest events in the history of the earth, are of the greatest interest to geologists.

There was a time when the earth was without life, a time when the first living germ heralded a new birth, the transformation of a world that was dead, and yet a world vibrant with physical energy, into a world endowed with endless potentialities which found expression in progressive development, concurrently with occasional retrogression as age succeeded age. How and when this miracle occurred is one of the secrets that will remain beyond human reach: we can only speculate on the manner of the transformation—the greatest event in the history of our planet—and allow our imagination to follow the creation of an invisible particle of living protoplasm in the water of a primaeval ocean. We think of germs vibrant with life, endowed with properties giving them the power of growth and reproduction: germs having within them a potentiality of infinite development into more and more complex organisms, culminating in man. We see in imagination some of the earliest descendants of ultra-microscopical particles of protoplasm, minute actively mobile units, neither plant nor animal, but members of a borderland kingdom, living examples of which are well known to biologists. It is not improbable that existing borderland organisms, with attributes common to the plant and animal kingdoms, may be descendants of pre-Cambrian ancestors, illustrating persistence of design throughout the span of geological history, types that survived as links with an early stage in evolution, emblems of continuity and persistence giving us a glimpse of immortality and of changelessness unaffected by the urge to progressive development. From a stage in the pre-Cambrian era until the beginning of the Cambrian period, when the seas were inhabited by myriads of animals of many kinds—not only simple forms of life, but others already relatively complex—the unfolding of life must have been a gradual upward progress from the infinitely small to forms in which complexity was an expression of division of labour, and increasing specialization of function. Between the Cambrian animals and the unknown ancestors of the groups to which they belong we are bound to assume a wide gap in time, sufficient for the operation of evolutionary processes, a gap which could be filled had the intervening forms been preserved. Unfortunately the oldest known sedimentary rocks tell us nothing of any real value of the long procession of living things, both animals and plants, which passed across the stage in the long interval separating the dawn of life from the early days of the Cambrian period when evolution had reached an astonishingly high level. Several fossils have been described from pre-Cambrian rocks, some no doubt the remains of animals of the sea, but they afford very little information, others are of very doubtful origin, possibly organic, and perhaps more probably inorganic structures simulating animal or plant forms. Beds of graphite, almost pure carbon, in the State of New York, in Finland and elsewhere, may be comparable in origin with seams of coal in a more advanced stage of alteration of plant material: they have yielded no recognizable fossils, but may perhaps be accepted as indirect evidence of plant life.

It would seem reasonable to suppose that when the earth was young the climate was much warmer than it is to-day, also that the wrinkling and folding of the crust decreased in violence as the loss of intense heat continued. We know little of the climatic conditions in the earliest periods, but we do know that boulder-beds, comparable with those of the last Glacial period, were discovered in the province of Ontario in Canada, a fact which makes us realize that even in the pre-Cambrian era the climate was not everywhere tropical. It was pointed out in an earlier chapter that the old view of a gradually cooling earth has been substantially modified since the discovery of radio-active elements from which fresh supplies of heat are constantly being released. While there is good reason to believe that the ebb and flow of tides was on a larger scale in the pre-Cambrian era, when the distance separating the moon from the earth was less than it subsequently became, we are not justified in assuming that on the whole the factors conditioning rock-destruction and rock-building were on a very different scale from those operating in the world as it is.

We pass now to a brief description of pre-Cambrian rocks in the British Isles, taking Scotland first, as it is there they occupy by far the largest area, and afford the fullest information. The greater part of the country north and north-west of the Great Glen Fault, from the Moray Firth to Loch Linnhe, is built of pre-Cambrian rocks; in a few districts, especially in the north-eastern corner, and south to the Moray Firth, Old Red Sand-stone rocks are exposed at the surface. The fiord-indented coastal strip from Loch Eriboll in the north to the Point of Sleat in the island of Skye was mentioned in Chapter v, p. 72; the rocks of this North-West Highland belt are pre-Cambrian, Cambrian and Ordovician. It is separated by north-to-south fractures from the rest of the Scottish mainland; the most important is known as the Moine thrust-plane, a line of fracture inclined at a low angle separating the North-West Highlands from the region to the east where the rocks consist of a complex mixture grouped together as the Moine Schists because layered glistening schists—products of intense metamorphic action—make up a large proportion of the whole. Similar rocks cover much of the Highland moorland between the Great Glen Fault and the approximately parallel Highland Boundary Fault which runs from Stonehaven, south of Aberdeen, to the Firth of Clyde and across the island of Arran. Many of the highest mountains of the Scottish Highlands are made of schists and other pre-Cambrian rocks. Schists are readily recognizable by the light-reflecting flakes of mica scattered over the surface, by the tendency to split into layers, the smoothness of the layers and the frequent bending of the thin slabs into small folds. These rocks are sediments which have acquired a crystalline structure through metamorphism; they are not igneous in origin. The precise age of the Eastern or Moine Schists and the accompanying rocks is still unsettled, the view in most favour is that they are highly altered pre-Cambrian sedimentary beds, though some may be Cambrian. Lack of fossils is one of the difficulties in the way of assigning the complex to a definite position in the geological table. No attempt will be made to discuss the keenly debated problems raised by the rocks which give to the Grampian Hills and other Highland districts the rugged beauty of the awe-inspiring scenery; they are not the oldest rocks in Scotland though many of them are most probably much older than those upraised from the Cambrian sea.

The foundation-stones of Scotland, that is, the oldest and lowest part of the crust accessible to us, are mainly gneisses, banded rocks similar in composition to granite. It is not always easy to distinguish a banded gneiss with its light and dark layers from a sedimentary rock made of material—quartz, felspar, and mica—derived from the wear and tear of granite and other crystalline rocks of igneous origin. Some rocks that have been called gneiss and believed to be igneous are, in fact, very highly altered beds of sediment. On the other hand, many gneisses were produced on the cooling and crystallization of molten material, their characteristic banded structure having been superinduced by the flow of the component minerals in response to pressure which converted a solid unbanded mass into a typical gneiss by rendering it plastic enough to admit of movement of the constituent crystals. Comparison has been made of the layered structure of gneiss with the wavy, curling lines of foam frequently seen on the surface of a pool at the foot of a waterfall, where gentle eddies leave their impress in the arrangements of the foam and bubbles.

Pre-Cambrian gneissic and granite rocks are the chief components of the Outer Hebrides, from Barra Head to the Butt of Lewis, the western outposts of Europe. Harris, Lewis and the smaller isles are the worn-down roots of a mountain range on a continent of which only dismembered fragments remain. Stepping from the deck of a steamer on to the island of Lewis, a visitor who knows nothing of geology misses the thrill experienced by those who read sermons in stones: the smooth contours of the rounded hillocks speak with no uncertain voice of the unmistakable finger-prints left by a thick overriding sheet of ice in the last Glacial period. Turning from the records of the Ice Age—an event of yesterday in the geological sense—to the solid rocks themselves, we are transported to the very beginning of the earliest chapter of earth-history of which any pages remain; we stand on the edge of a continent now represented by partially submerged remains of what was once a far-flung land, a land that was in all probability untenanted by any living things, an empty stage set for the drama of life. The Outer Hebridean rocks are practically the same as those which compose the main part of the Laurentian shield on the other side of the Atlantic. Ice-worn blocks of the same North American shield are conspicuous objects in Central Park in the heart of New York City. At some localities on the shores of Disko Island, off the west coast of Greenland, and over the whole of Finland, precisely the same coarsely crystalline rocks are exposed.

Traversing the deep barrier of sea on the east of the Outer Hebrides, the same Pre-Cambrian rocks are seen in Iona, Coll, Tiree, Islay and other islands of the Inner Hebrides. Part of the island of Skye is made of the same rocks, and they bring us into touch with the larger area in the North-West Highlands, where they can best be studied more closely. It will be convenient to consider together two groups of the oldest known rocks: a group in which gneiss plays a dominant part, gneiss that has been called Lewisian from the island of Lewis; also a younger group known as the Torridonian series, because it is well displayed in the Loch

Torridon district. Both these groups have been most thoroughly investigated in the coastal strip from Cape Wrath and Loch Eriboll to Skye, bounded on the east by the great Moine thrust-plane. The Lewisian gneiss is seen in the high vertical face of Cape Wrath, where dark bands of an intruded finer-grained rock are a conspicuous feature: this gneiss is the rock of the rounded downs and the ridges from Cape Wrath to Loch Torridon; it is seen also in the Loch Maree district, at Lochinver, and many other localities on the mainland, also in the islands of Rona and Raasay. The Torridonian series is entirely different in origin; it is made up of several thousand feet of coarse and fine-grained layers of sedimentary material, in contrast to the crystalline Lewisian gneiss. The Torridonian rocks are widespread in the North-West Highlands; they are the masonry of many mountains from the north coast to Gairloch, Applecross and Skye. It is these rocks that are responsible for the wild mountain scenery south from Loch Maree. At several places in the North-West Highlands, the two groups of rocks—the Lewisian and the Torridonian—are seen in contact: the gneiss forming hills and valleys of the oldest land in the British Isles, the Torridonian beds lying in horizontal sheets on the irregular, eroded surface of the gneiss. Though the two series are in contact they were separated in time by a long interval during which many scenes were enacted. At one time enormous piles of Torridonian rocks were spread over the whole area, completely hiding the much more ancient Lewisian gneiss; their present occurrence on isolated hills, separated by valleys, is the result of the gradual wearing away of a once continuous plateau, of which they were the upper courses. The Torridonian rocks are mainly composed of an aggregate of material derived from granitic and gneissic rocks by the eroding action of agents of destruction, mostly large and small pieces of quartz and fragments of felspar; they are sediments that were probably deposited in shallow water, and no doubt in part on a land-surface. There is evidence that at least during part of the Torridonian age the climate was cold and dry; the fresh condition of the felspar suggests lack of moisture which would have hastened decay and disintegration: pebbles have been found which have smoothed flat faces meeting at a clean-cut ridge, a form familiar in wind-swept deserts. Pieces of Torridonian rock might easily be mistaken for a rather fine-grained granite, which they super-ficially resemble, and this is not surprising; the sedimentary beds are made of the debris of granitic rocks. More careful examina-tion shows that the grains of quartz and other minerals do not form a crystalline mass in which the particles form a connected dovetailed complex, as in rocks of igneous origin, but an aggre-gate of separate units. There are also conglomerates of water-worn pebbles, breccias of angular stones comparable with screes or other rock debris formed on land and not in water. With these are beds of shale, muddy sediment of rather deeper water. Pits on the surface of the finer-grained layers are evidence of rain-storms, and cracks due to shrinkage suggest muddy beaches exposed to strong sunlight. The Torridonian rocks were de-posited from rivers, at times swollen by flood-water, at times flowing with reduced momentum, sediment spread over sub-merged valleys and around the hills of Lewisian gneiss, until they covered the whole region. As the years passed, the land rose above water-level and rock-destruction took the place of rock-building; the ground was carved into valleys and hills, as the path of rivers was determined by lines of weakness and least resistance. Thus was fashioned out of the upraised tableland the awesome mountain grandeur of the North-West Highlands. The removal of the horizontal sheets of Torridonian rock laid bare portions of the old Lewisian landscape as it was before submergence and the deposition of the detritus spread as a concealing mantle over the buried gneiss. Gazing at the rounded, weather-beaten hills of Lewisian rocks, still partially enfolded by the Torridonian covering on the flanks of Quinag (2653 ft.) and other Highland mountains, we are face to face with a scene that was typical of a very early stage of the Pre-Cambrian era; we are privileged to see revealed the reconstruction of a land-scape taking us back through at least a thousand million years, to the oldest known land in the British Isles.

The upper part of Quinag and summits of other hills are made of a cap of white Cambrian rock, a metamor-phosed sandy sediment which was deposited over the Torri-donian beds of coarser texture, after they had sunk below the waters of an encroaching sea. Reference has already been made to the gigantic scale of the physical forces which caused devastation and confusion in the rocks of the North-West Highlands after the Cambrian period. Thousands of feet of Lewisian, Torridonian, and Cambrian rocks were involved in the debacle; huge portions of the crust were rent asunder and moved bodily along lines of fracture a distance of many miles, eventually coming to rest as overthrust blocks, older rocks lying on younger rocks. Slices of Lewisian gneiss, more than 1000 ft. in thickness, now rest on overturned sheets of Torridonian sediments. Beyond the Moine thrust-plane, on the eastern border of the western coastal strip, the eastern schists beneath the moorlands of the Central Highlands were also implicated in this terrific storm among the rocks; their complex structure has not yet been fully interpreted, and we leave them as a problem that will long continue to stimulate the ingenuity of geologists.

Pre-Cambrian rocks, gneiss, volcanic lavas and ashes, and sedimentary beds occupy a large area in Anglesey; they are exposed also in Holyhead mountain (720 ft.) and form a rocky rise from Bangor to Caernarvon. Rocks of the same era are seen in the Wrekin, and in the cliffs near St David's in South

Wales. They build the higher ground of Charnwood Forest; they occur at Nuneaton, the Malvern and Lickey Hills; in the Lizard peninsula, Kynance Cove, and other localities in Cornwall. The precise age of the rocks at Kynance Cove, and other places on the wildly beautiful Cornish coast, has not been definitely settled; they are certainly early Palaeozoic, and probably in part pre-Cambrian. For the most part pre-Cambrian rocks of England are buried below sediments of later periods, where they lie far beyond our reach as the continuation of the Archaean platform which was the foundation of the whole superstructure of the world.

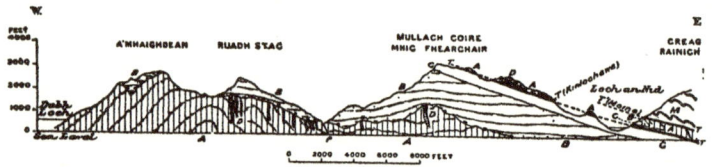

Fig. 10. Section of pre-Cambrian and Cambrian rocks in the North-West Highlands of Scotland. *A*, pre-Cambrian gneiss; *B*, Torridonian beds; *C*, Cambrian strata; *D*, dykes in the gneiss; *M*, Moine schists; *TT'*, thrust-planes; *f*, fault.

The end of our journey is reached: passing hurriedly through a vertical section of the earth's crust many miles in depth, sediments raised from the floors of seas, estuaries and fresh-water lakes, beds of lava and volcanic ash, great mounds of granite and other rocks, slowly solidified from melted matter which had welled upwards from subterranean sources; all these and many other kinds are the components of the mosaic of which the land surface is made. The angles of inclination of rocks, their varying power of resistance to Nature's cutting-tools, are factors which have helped to make the British region famous for its unsurpassed range and diversity of scenery. As we descended from more recent to successively older layers of the crust, glimpses were obtained of the moving panorama of life on land and in the sea, of the age-long procession of changing companies of actors. At length we reach the longest and the oldest chapter of the earth's history, we are in a world where life began, a world in its physical environment, in its relation to the forces of Nature differing but little from that in which we live, a world warmed by the sun, under the same canopy of sky by day and the same star-lit dome at night.

'That which hath been is now.' We have voyaged over strange seas, gathered scraps from the litter of woodland and forest that gave colour to changing landscapes; we have explored 'the stormy bases of the world' and 'the dust of continents to be'. To our vision has been given a greater power of penetration; we have read a little of the Epic of Creation and, little though it is, we have been brought nearer to a perception of the infinite.

CHAPTER XVII

The Procession of Life

References in the foregoing chapters to a few animals and plants discovered in rocks of different periods barely touch the fringe of a large and important section of geological history. In this chapter some repetition is inevitable: my aim is to give a general account illustrating the kind of information furnished by fossils and its bearing upon the history of life and the problem of evolution, so far as that is possible in a few pages. The history of the earth embraces a survey of physical as well as organic evolution. As Field-Marshal Smuts reminds us, the difference between the physical and organic worlds is a difference between two kinds of activity: intensely active matter and life overlap. The ambition of geologists is to reconstruct the background at successive stages in the earth's development, and the procession of life through the ages. An adequate treatment of the procession of life as revealed by fossils would involve the co-operation of many specialists and excursions into technical byways of botanical and zoological science. All that is attempted in the following pages is to make acquaintance with a selected number of the ghostly companions encountered by those who endeavour to follow the onward sweep of life as recorded in the depths of the earth, and not least to emphasize the fact that 'Nature in all its richness' can only be appreciated when to knowledge of the present are added the memorials of the past. Fossils hammered out of a rock attract us by their antiquity; they speak to us as messengers from seas and lands of other days, and help us to decipher Nature's manuscripts. Readers wishing to know more of an exceptionally fascinating and fruitful subject may perhaps be persuaded to extend their enquiry by reading books on palaeontology and, it is hoped, follow the example of many amateurs to whom geology owes a great debt, and apply themselves to an intensive study of a selected family or genus of animals or plants.

In many classes of animals, and in certain classes of plants, extinct types exceed in variety and number those that still exist. In order to interpret fossils, and assign them as far as possible to their respective places in the animal and plant kingdoms, the first essential is to study the living, in order to be qualified to distinguish between extinct types which cannot be accommodated in any subdivision of the two kingdoms based exclusively on living forms, and those that can be matched with related types in the present. The publication of the *Origin of Species* in 1859 gave a tremendous impulse to exploration of the rocks in search of evidence of the truth of Darwin's theory. Facts relevant to the problem of evolution are obtained both from the study of living organisms, and more especially from animals and plants of former ages. One result of the revolution in thought caused by Darwin's book was a recognition of the importance of a more thorough knowledge not only of adult animals and plants, but also of their embryology, that is, the branch of biology concerned with the stages through which an individual passes in the course of its development from a single cell—the fertilized egg—to maturity. Embryological research has shown that all the vertebrate animals (backboned animals), though differing widely in the adult state, have more in common in the earlier stages of their development: embryonic characters which are transitory, and no longer present in the adult, have been recognized in fully developed extinct animals, a fact pointing to the survival in the embryos of existing genera of primitive features characteristic of mature animals of former epochs. One example of an embryological character having a

SG   18 direct bearing upon relationship is the discovery of gill-slits or traces of slits in the immature forms of all classes of vertebrates, an indication that certain animals, now without gills in the adult state, were evolved from aquatic ancestors. Another instance of light thrown by embryology on descent is furnished by birds: several extinct birds which lived in the Mesozoic era had well-developed teeth; living birds have none, but in embryo parrots and ostriches faint dental ridges seen in the embryos are vestiges of a character possessed by extinct birds and now only reflected in a transitory survival or memory of the past. It is, however, to fossils, to remains of extinct types, that the final appeal for evidence must be made.

The classification of animals and plants given in text-books is based upon existing forms of life: the ultimate aim of botanists and zoologists is to draw up classifications which express relationships and approach as close as possible to the ideal of natural classifications, in contrast to artificial systems that serve primarily as aids to the recognition of the several families, genera and species. Mammals, the highest and most highly specialized animals and the present dominant class, amphibia, reptiles, birds, fishes, and other classes are characterized by well-defined, distinctive features, and it is not surprising that they used to be regarded as so many unconnected and separate creations. After general acceptance of the doctrine of evolution as a working hypothesis, attention was focused upon a search for connecting links between one class and another. The search for truth—cynically defined as the hypothesis which works best—was stimulated. Missing links are still being sought; a few, but only a few, have been found. Were it possible to restore the past, to have before us a representative collection of the animals and plants of all time, we should be in a position to read Nature's secrets and to follow the unfolding of life from one age to another. This ideal will never be reached. One of the greatest obstacles confronting searchers after connecting links is, as Darwin fully realized, the imperfection of the geological record. Knowing something of the physical history of the earth, recurrent convulsions, cycles of mountain-building and flooding of the land by transgressing seas, we are compelled to admit that there must have been wholesale destruction of the records of ancient life. The fossils available to us are, of necessity, a very small fraction even of such animals and plants as possessed coverings or hard parts favouring preservation. Most fossils found in marine sediments are shells of the calcareous skeletal framework which furnish partial clues to the nature of the soft parts, though of the actual bodies seldom anything remains. As one would expect, impressions left by soft-bodied animals on rock-surfaces are very rare: among the few that are known, a clearly defined impression of a jelly-fish on a rock of Lower Palaeozoic age is a striking example. Remains of land animals are often preserved in abundance and in good condition, but apart from the bones there is little else. Turning to plants, it is only when their remains have been petrified that much can be learnt of their anatomical characters; the vast majority are preserved with little or no structure. Rocks contain many undiscovered treasures which future search may bring to light, but there will always be rocks beyond our reach and there will always be gaps in the story which remain unfilled. What then has been contributed by fossils to a better understanding of the history of life?

One of the most interesting results achieved by an examination of Nature's zoological museum and the herbaria of the rocks is the discovery of many animals and plants of a type now unknown, members of extinct classes, groups and families. Some are transitional or, more correctly, generalized types in which are combined, in a single animal or plant, characters no longer met with in association but distributed among genera or species belonging to different groups. These generalized types in some instances are accepted as evidence of the common origin of groups which now retain no such relics of descent from one ancestral stock. Some of them point the way to progressive development; others lead only to a cul-de-sac where they are 'wiped out of the book of the living'.

Reference was made in an earlier chapter to fossil skulls and other bones which lend support to the derivation of *Homo sapiens* from ape-like ancestors: bones of early man are rare, partly, no doubt, because in those days man was a very rare animal. Fossils of flesh-eating (carnivorous) animals, such as tigers, lions, hyaenas, and many others, afford some indication of a possible link between early carnivores and whales. One of the most familiar instances of fossils throwing much more light on the course of evolution than could be obtained from living animals alone is the series of extinct Tertiary genera leading up through earlier forms to the modern horse, showing a gradual reduction in the number of toes from the primitive Eocene horse, a relatively very small creature with five well-developed toes, through intermediate types culminating in the present-day single-toed animal which still possesses two additional, small and useless splint bones, vestiges reminiscent of a three-toed ancestor. Fossil bones have furnished much information on the early history of elephants, camels, the hippopotamus and other vertebrates: they show that these large animals had comparatively diminutive precursors; it has been possible to trace the successive steps by which elephants acquired their trunk—not by the method ingeniously envisaged by Kipling—and other distinguishing features. The modern South American sloth, a peculiarly lethargic beast apparently content to spend most of its time between meals suspended upside down on the branch of a tree, had forbears such as the extinct giant sloth (*Mega-therium*) from the Quaternary Pampas of South America, a giant which lived on the ground, supporting itself on massive hind-legs and a substantial tail and using its short fore-limbs as arms with which to reach the boughs of tall trees.

Much light has been thrown by fossils on the past history and evolution of birds. The famous *Archaeopteryx* was discovered many years ago in the fine-grained Lithographic Slate of Bavaria, the stone used by lithographers, a sedimentary bed of Jurassic age; this creature, about the size of a pigeon, had feathered wings and a lizard-like tail with feathers; it had also teeth. *Archaeopteryx* was probably incapable of flight and used its wings for gliding from trees to catch fish in the water below. It has been pointed out that the young of the hoatzin, a primitive bird of British Guiana, climbs trees from which it swoops upon its prey. The Jurassic genus is particularly valuable as evidence of relationship between birds and reptiles. The *Apteryx* (kiwi) of New Zealand is an example of a living winged but flightless bird, also the ostrich.

The class Mammalia, including man and the largest living animals, occupies a position in the animal kingdom corresponding to that of flowering plants in the plant kingdom. A lower jaw with teeth, belonging to a small animal discovered in the Jurassic Stonesfield Slate of Oxfordshire, is one of the earliest examples of the class, an advanced guard of a race that was destined to achieve domination in the Tertiary era. It has also been proved that reptiles, now relatively small, were preceded by a very large and varied company of animals which swam in Jurassic and Cretaceous seas. The most ancient reptilian remains are from rocks of the Permian period. One of the most interesting examples of an extinct Cretaceous and Jurassic reptile is the well-known *Pterodactyl*, the finger-winged creature (see Chapter x, p. 180); it was like a large bat in having wings that were not made of feathers. Some of these winged creatures were about the size of a crow, others were larger; they were gliders and did not fly by flapping the wings. As John Thorneley of Cambridge says:

Though it could not boast of beauty, and it did not care to sing,
In the eye of evolution it's an interesting thing.

A remarkable instance of contrasts between past and present animals is furnished by the discovery in several parts of the world of a varied assortment of Dinosaurs, extinct members of the class Reptilia (Chapter x). They have been traced as far back as the Permian period and died out in the Cretaceous; they reached their greatest development in size and number in the Jurassic and Cretaceous periods. Some of the earliest Dinosaurs were small, little larger than a cat, but in the course of the latter part of the Mesozoic era they surpassed in size all other animals of the land, both recent and extinct. The genus *Diplodocus* (that is, double-beam, so called because of two bones which protected the blood-vessels in its hind-quarters when the long tail was being dragged along the ground) was a giant over 80 ft. long; like most of its kind it was a vegetarian. The largest flesh-eating Dinosaur, *Tyrannosaurus*, discovered with several other genera in Cretaceous rocks in North America, was nearly 50 ft. in length; its head over 4 ft. in length, its jaws set with sharp-edged teeth. In size and shape resembling mammals, they were structurally reptilian, and it is interesting to note they possessed characters now found in both birds and mammals. Many were armour-plated like mechanized cavalry, provided with large shield-like excrescences and horns comparable to the very much smaller hooded outgrowths on some existing scaly lizards. In addition to skeletons, a few mummified specimens showing the actual skin and a number of eggs have been discovered. Dinosaurs wallowed in swamps, or pottered clumsily over land; huge in body, they had ridiculously small brains: a dinosaur heavier than an elephant had a brain esti-

mated to weigh 2½ oz., approximately that of a three-weeks-old
kitten. The reptile *Iguanodon*, mentioned in an earlier chapter
(Chapter x), was discovered more than a century ago in a
Lower Cretaceous sedimentary bed in Sussex. These and other
extinct reptiles, which disappeared before the Tertiary era, had
been able to wander over a vast region of the world, from
Scotland to South Africa and the American continent; their
failure to survive was, no doubt, in part due to their inconvenient
bulk and the wholly disproportionate development of brain.
Thorneley's uncomplimentary tribute to these monsters is something more than a poet's fancy:

There were large and lumpish lizards in that old and foolish
  time,
Which led a dull existence in a sort of slushy slime,

.   .   .   .

With the maximum of body went the minimum of mind,
When the rest were romping forward, they came rumbling
  on behind.
They may pass as early efforts, but they did not run to brain,
So Nature swept them all away and set to work again.

  In the swamps and lagoons of the Coal Age the most highly
developed backboned animals were related to existing salamanders, members of the class Amphibia. Some of them were
covered with hard armour-like plates like crocodiles and alli-
gators, but they were not reptiles; these had not yet been evolved.
There were no birds in the forests: the only flying creatures
were insects such as dragonflies, and cockroaches crawled among
the herbage—forms distantly related to living species. The oldest
fishes are from Ordovician and Silurian rocks, genera with a
cartilaginous instead of the jointed bony backbone of late types,
a tough outer skin with knobs and protective plates, some with
large and broad swimming paddles in place of the more efficient
fins of modern fishes, which swam in Jurassic and later seas.

Some Palaeozoic genera, though bearing little resemblance to the ordinary living genera, are remotely connected in certain characters with the most primitive examples of the class that still exist; the Port Jackson shark of Australian waters (*Cestracion*), the mud-fish (*Ceratodus*) of Queensland, *Lepidosiren* of South America, and *Protopterus* of African lakes are surviving relics of groups that were abundantly represented in the Old Red Sandstone lakes.

What, it may be asked, is the scientific value of the facts gleaned from fossils? Many fossils confirm suspicions of relationship between sections of the animal world based on a study of living forms and furnish a convincing evidence of affinity; they strengthen belief in the fundamental principle of evolution. In contrast to the isolated and rare examples of links with the past which have persisted to the present day, the diversity in size, in design and external appearance of the numerous extinct types enable us to picture animals of far-off ages in the heyday of their vigour when they flourished over wide regions on land and in the sea. Diversity was shown by the range in form and structure, an expression of rapidity of development of species after species, an unconscious striving after an ideal that was never attained. In these different types, different and yet related, we see examples of Nature's experiments, a groping after something higher and more permanent, which seem to us to have been failures. There are, however, indications that from some of these apparent failures were evolved new plans of construction which, in course of time, were firmly established and, as progress continued, became the basic characters of a new and more successful line, for example the evolution of mammals which ultimately ousted Dinosaurs from their pride of place.

A similar chronicle of success and failure is furnished by animals belonging to the Mollusca, one of the subdivisions of the Invertebrata, that is, lacking a backbone, soft-bodied creatures of sea and land, most of them protected by shells. Mention has previously been made of the past history of *Nautilus*, a genus still living in the Indian and Pacific Oceans; also to the wholly extinct Ammonites, with spirally constructed shells, in some forms more than 6 ft. in diameter. The great variety and abundance of Ammonites in Jurassic rocks—the period in which they reached the peak in development and distribution in world seas—is proof of rapidity of the evolution of new forms. There is reason to believe that the spiral construction of the chambered shells of *Nautilus* and the Ammonites was preceded by an earlier type in which the chambers were arranged one above the other within a slightly bent or perfectly straight containing wall: the straight forms are common fossils in earlier Palaeozoic rocks.

A passing reference was made in an earlier chapter to two classes of animals in which the soft body is enclosed in two closely fitting shells, the much larger class—Lamellibranchs (plated gills), including oysters, mussels, cockles and many more common bivalves, and the much smaller and quite unrelated class, the Brachiopods, often called lamp shells because of a similarity in form, in side-view, to an old Roman lamp. Though superficially not dissimilar, the two classes are distinguished by several well-marked features. In a lamellibranch the two valves are as a rule substantially equal, they are right and left in relation to the body and there is no aperture in the shell for a stalk of attachment to the sea-floor. In a brachiopod the two valves are obviously unequal, one projecting in the apical region over the other and perforated by an aperture to allow the prolongation of part of the body as a stalk anchoring the animal to mud or sand. Brachiopods are exclusively marine; some lamellibranchs live in fresh water, e.g. the fresh-water mussel. The two groups illustrate an interesting reversal in abundance when we compare their histories: there are now about 160 species of brachiopods and, among them, the genus *Lingula* which burrows into the floor of shallow seas. *Lingula* is a striking example of persistence and conservation: this genus, now comparatively rare, lived in Ordovician seas and has continued until the present age with little change. In ancient seas, brachiopods played a prominent part and were represented by hundreds of species and many genera; on the other hand, lamellibranchs, now very common and widely distributed, were formerly much less abundant than brachiopods. The position of the two classes has been reversed. A notable example of a persistent lamellibranch is the genus *Trigonia* previously mentioned as a common Jurassic bivalve; it is still found off the coasts of Australia. Another conservative genus, not a bivalve but an animal with a single spirally grooved and tapered shell, is *Pleurotomaria*, common in early Jurassic seas, less common in Silurian seas, and represented by a few living species. As Dr Bather said: '*Pleurotomaria* seems to have found a nook in creation which just fits it.' These and other survivals, often rare and bordering on extinction, are like echoes from vanished worlds.

The study of extinct members of the large class Arthropoda, including insects, spiders, scorpions, lobsters and other groups, has revealed many contrasts between the present and the past. Eurypterids, briefly described in an earlier chapter (Chapter XIV), have not been recorded from rocks later than the Carboniferous period; they were plentiful in Old Red Sandstone lakes and have been traced as far back as the Cambrian period; some of them were as much as 9 to 10 ft. long, strange-looking creatures with an armoured skin: they may be compared structurally with scorpions, but they lived in water. King crabs (*Limulus*) are not related to crabs; they may be described as a kind of aquatic scorpion, which, in appearance and habits, are among living animals the nearest to eurypterids and trilobites. Trilobites have already been described (see Chapter xv); these early crustacea, crawlers on sea-floors, were represented by more than 300 species in Cambrian seas and some of them were more than a foot long—the largest of their kind; they occur in the older Palaeozoic rocks all over the world.

Two marine animals familiar to those among us who pick up shells on sea-beaches and look into rock-pools—the flat-bottomed globular shells of sea-urchins made of radially disposed calcareous plates, and the soft-bodied colourful sea-anemones—are examples of two classes that have existed through many stages of geological history. Sea-urchins, as we have seen, left many well-preserved shells and detached spines in the English Chalk: there is not much difference in the general plan of construction between genera from Jurassic and Cretaceous rocks and those that live in modern seas. The group is one of many that lived vigorously through many ages, continuing to produce new forms and showing little sign of decadence. Sea-anemones are ill-adapted for preservation as fossils, but their near relations, the coral-building polyps, which are widely spread in the warmer oceans where reefs fringe the coast and in deeper water encircle the blue lagoons of coral islands, are some of the most abundant and readily recognizable fossils. Limestones of the Carboniferous, Devonian and Silurian periods are rich in corals, both single cups and compact masses, broken by waves from Palaeozoic reefs. They, like sea-urchins, illustrate persistence, power to endure and to produce new types through successive epochs representing at least 500 million years.

A less familiar subdivision of the large class which includes sea-urchins and starfishes is now represented mainly in tropical seas by the so-called sea-lilies or crinoids; most of them are attached to the sea-floor by a stalk, sometimes several feet long, made of a series of small calcareous joints or plates supporting at the top an oval 'calyx' containing the body and encircled at the upper end by numerous arm-like tentacles, chains of smaller limy segments. These animals are exceptionally attractive fossils and are striking objects on slabs of Jurassic sedimentary beds, on which they often form tangled masses of intertwined slender stalks ('encrinites') anchored to the mud and bearing the swollen body-cases at the tapered summit. The sea-lilies are closely allied to starfishes and there are now about a dozen kinds in deep and warm seas: their ancestry has been traced to the Cambrian period. They were abundant in Palaeozoic and especially in Liassic seas. A Mediterranean genus, the feather star (*Comatula*) which occurs also on the rocky shores of France and England, has no stalk in the adult state, but the young *Comatula* is provided with a slender stalk which, as growth proceeds, gradually shrinks; the mature animal is a sea-lily which has lost its stalk. It is interesting to recall that the first sea-lilies discovered were Jurassic fossils and it was not until later that the first living specimens were found off the island of Martinique. Fossils were known before the discovery of living descendants.

Enough has been said to give a general idea of the kind of information furnished by fossil animals. As the several classes are followed through the aeons of geological history great differences become apparent: many classes have existed without more than minor modifications since the Palaeozoic era; others, now reduced to a very small company, were formerly much more vigorous and more widely scattered; other classes, of which there is no memorial in the present age, played a leading part on land or in the sea in the Mesozoic or Palaeozoic era. One outstanding fact is clearly demonstrated: travelling backwards we see the mammals increasing in number and variety in the course of the Tertiary era, and as we descend deeper the only examples of this now dominant class are a few rare fossils in rocks of the Triassic and Jurassic periods, small creatures which it is difficult to realize were the forerunners of the hosts that were to come. The reptiles of the modern world are the relatively small and degraded relics of a much greater company that had its roots in the Permian period. Existing amphibia had ancestors in the forests of the Coal Age. Each of these classes reached its highest level, as measured by range in form and abundance of genera and species, at successive periods of the history of the earth. This we know, but despite the wealth of material collected from the rocks of many ages, we know comparatively little of the steps by which these and other classes of the animal kingdom were evolved, the characteristics of their remote forbears.

The records of the rocks, taken as a whole, provide an amazing picture of the moving panorama of life, the rise and fall of many dynasties and the endless fertility of the womb of Nature; they increase our faith in the doctrine first enunciated in its fullness by Darwin, though, when all is said, we are conscious of the little that is known of the manner, the *modus operandi*, of the forces and influences that governed and conditioned the unfolding of life. Attention was called in Chapter xv to a most important discovery, several years ago, by an American geologist in the Rocky Mountains of a remarkable assemblage of several hundred sorts of Cambrian marine fossils, soft-bodied animals and others provided with a protective covering of chitin or some other resistant material. Within an area of 40 sq. ft., in beds of shale 7 ft. in thickness, 56 different genera were identified. Genera and species of practically all the known classes of invertebrates were found in a good state of preservation, and a significant fact is that these Cambrian types are not by any means all primitive and simple, but complex and far advanced towards specialization and having the attributes of organisms in which division of labour had already attained full expression. How far back beyond the limits of the Palaeozoic era the chain of life continued we do not know; we only know that if, as we believe, evolution was a gradual process and organisms were not created in their completeness, the ancestors of the Cambrian animals must have lived earlier still in the pre-Cambrian era and left no memorial. It used to be fashionable to construct genealogical trees illustrating in graphic form how the various classes were supposed to have arisen as branches of a common stock; these trees were rather hypothetical than factual. Another picture of evolution, and probably closer to reality, represents most classes as separate and more or less parallel lines, each starting from some hypothetical simple beginning, and broadening and branching as they traverse the geological series, bundles of lines and not branches of a common trunk. The story of evolution is not as simple as it has often been described; it is a record of fluctuation, of rapid ascents from older types small in size and few in number to a multitude including monstrous animals of gigantic size that seem to be examples of Nature's efforts that went awry. Then, as time passed, there followed an equally rapid descent often leading to total eclipse, sometimes to a few diminutive survivals, all that remain to remind us of the glories of old time.

Leaves, stems, fruits, and seeds and other scraps from forests of many ages, unconsidered trifles with others worthy to be called jewels of the rocks because of the light they have thrown on the history of plant life, have helped substantially to increase our vision and taught us to recognize that in many respects the past is the key to the present. How did life on this planet begin? At what stage in the history of the earth were the foundations of the living world laid?—are questions which have long quickened the imagination and stirred the curiosity of biologists. Another question put to Nature is—which are older, animals or plants? Can we follow the march of evolution of the plant kingdom, see the steps by which one class emerged from another or as a separate line, and recognize among the large number of extinct classes and groups examples of connecting links, transitional forms of which there is little or no evidence in the plant world of the present? To the first question—the origin of life— we can never hope to give an answer supported by observed facts. The desire to see beyond the most distant horizon to heights invisible to human eyes is a natural ambition and an incentive to man's highest endeavour. Though a satisfying answer to the second question may be beyond our power, biological knowledge enables us dimly to visualize the course of events in the earlier stages of life's upward trend. The oldest living organisms did not possess the distinctive features now associated with animals and plants respectively. Certain very simple microscopic motile organisms are assigned to a class apart and occupy a sort of neutral territory, a borderland between the two kingdoms. These primitive aquatic organisms may be described as links between the present and an inconceivably distant past when they were the vanguard of an army equipped with boundless potentialities. The third question covers several aspects of evolution which, if fully considered, would lead us far beyond the reasonable limits of a single chapter. Mention was made in an earlier chapter of the origin of the present dominant class in the plant kingdom, including the most highly specialized and most widely spread trees, shrubs and herbaceous plants ranging from the smallest herbs to the giant gum-trees (*Eucalyptus*) of Australia. Flowering plants, as this class is called by botanists, are represented by many fossils in Tertiary and in nearly all Cretaceous sedimentary rocks. They probably secured their present leading position in the first half of the Cretaceous period when many of the oldest known species differed very little from existing types.

The arrangement of families in modern systems of classification of flowering plants is intended to be an expression of their relative positions in order of increasing complexity and elaboration of structure; in other words, in the order of their evolution. Evidence supplied by extinct plants has obviously great value in confirming or correcting conclusions based solely on existing forms. It is therefore important to discover whether or not the oldest known flowering plants conform in such characters as are found in them to those possessed by living species believed to be relatively primitive. It is not possible definitely to assert that any one genus is the most ancient of its class, but it can at least be said that the history of the genus *Cercidiphyllum* has been traced as far into the past as that of any other flowering plant. *Cercidiphyllum* is a tree widely distributed in Japan and in mountain valleys in some parts of China: specimens may be seen in some English gardens and arboreta where the tree is grown for the sake of its most attractive autumnal colouring. This Far Eastern tree, now restricted to Japan and China, lived in early Cretaceous forests in North America, Europe and some hundreds of miles north of the Arctic Circle. Exceptionally well-preserved leaves, in addition to a few fruits and the characteristic small winged seeds, have been found in Eocene (early Tertiary) beds in the island of Mull as well as in Tertiary deposits in North America. *Cercidiphyllum* is one of a small number of flowering plants obtained from the oldest rocks of the Cretaceous period. It is usually regarded by botanists as one of the more primitive genera and is placed in a special family by itself because of its isolated position in the present vegetation of the world. Here there is a double reason for speaking of *Cercidiphyllum* as the most impressive link with the past—its proved antiquity and its primitive features which distinguish it from other living genera. It is one of many trees, now confined to the Far East, which once ranged over a far-flung territory in North America, Europe and Arctic regions. In the light of these facts, the brilliance of the leaves at the fall of the year may almost be described as a reflection of the golden age of a family now reduced to a single type, but millions of years before the birth of the human race represented by several closely related species in the western world.

The genus *Magnolia*, widely distributed on two sides of the Pacific Ocean, along the Himalayas, in China, Japan, Malaya, in Central and North America, is another primitive type which, though not as yet proved to be as old as *Cercidiphyllum*, flourished in Cretaceous and Tertiary forests in Arctic and temperate regions, in North America and Europe. The extinct species of this and other genera do not provide clues to the source whence they sprang: their ancestral history is one of many unsolved problems, but one positive result of researches into the geological history of plants is a greatly enlarged view and a better under-standing of geographical distribution. Many plants are confined to a comparatively restricted area on the present surface of the earth; others are known to occur over wide spaces in both the Northern and the Southern Hemispheres, and the large size of their territory might be interpreted as evidence of greater antiquity because the longer a genus or species has existed the more opportunity there has been for wandering over an in-creasingly large area. This view has been strongly advocated and supported by a formidable array of statistics, but when the facts of present distributional areas are supplemented by data gathered from the rocks, it becomes apparent that the present range of plants is, by itself, untrustworthy as a guide to age. Restricted distribution is in many instances a mark of antiquity, and of loss of vigour, in contrast to greater vigour, and a much wider area of occupation, in former periods, as we have seen in the history of *Cercidiphyllum*, which is only one of many similar examples illustrating the importance of viewing present distribution in the light of the past. Additional examples are supplied by several conifers and ferns: the redwoods and mammoth trees (*Sequoia*) of California, the Malayan fern *Matonia*, the Malayan and Indian fern *Dipteris*, which in recent years has at last been persuaded to grow in the fern-house at Kew, are all plants now confined within narrow geographical boundaries, but formerly almost cosmopolitan.

Belief in evolution carries with it acceptance of Darwin's view that each species of animal and plant began its life at one place, the place of its birth, and thence, in course of time, occupied a gradually enlarging territory. In other words, we believe in single centres of creation, and not in the sudden appearance of the same species in widely separated places. The late Professor Huxley put the case against special creation with his accustomed incisiveness: he said, 'If you believe in special creation, you must not be surprised if looking out of your window some morning, you see a brand new animal sitting on the lawn'. When the same species of trilobite, graptolite, or other animal of the sea is recorded from regions as remote from one another as North America, Europe, and Australia, we think of long voyages made in some instances partly by the animals themselves, and always with the help of ocean currents, and we are thus able to grasp the implication of tremendous periods of time. But with sedentary plants the position is very different. Some flowering plants, ferns and club-mosses are known to occur in Arctic regions, in the temperate zone of the North and South Hemispheres, and we express this fact by speaking of their migration over vast spaces. Migration, when applied to plants lacking the power of independent movement, may be regarded as roughly analogous to hitch-hiking; they are conveyed by external agents, wind carrying seeds and fruits provided with wings, or well adapted to aerial transport by their small size and lightness; birds and other animals consciously or unconsciously supplying means of transport; ocean currents and other aids to travel. Information on rates of migration is still meagre, and additional data, not very difficult to obtain, would be welcome. Some years ago a botanist made observations on the rate of travel of the fruits of a tropical tree (*Shorea*): this tree does not produce flowers until it·has reached a height of about 30 ft., and this is in approximately 30 years. The fruits have three large wings, and can be dispersed by wind as they slowly fall: some were found 100 yards from the parent tree. If any of them germinated in the new locality, it would be 30 years before the next stage in the journey could be undertaken: this means that *Shorea*, in favourable circumstances, could travel 300 yards in 100 years, or 100 miles in 58,666 years. The winged seeds of a Scots pine (*Pinus sylvestris*) have been picked up at a distance of more than 800 yards from the parent tree, and, assuming that the seeds were still able to germinate, this is an example of more rapid travelling. Migration measured by our limited standards is at best an extremely slow process.

Knowledge of ancient floras has raised several very difficult problems connected with climatic contrasts between the present and the past. Two examples must suffice: in the early part of the Tertiary era, the vegetation in Arctic regions, as far north as lat. 80° N., was very different from that at the present day. In place of arctic herbaceous plants and a few stunted willows and birches, there were forests in which many trees, including oaks, planes, *Cercidiphyllum*, *Magnolia*, several conifers, ferns, and other plants, played a prominent part.

One of the most impressive instances of the difficult and, it would appear, almost insoluble problems raised by fossil plants in relation to climatic conditions in the past has been furnished by Professor T. M. Harris of Reading. The facts are briefly as follows: several years ago a large collection of fossil plants was described from Rhaetic rocks in Scania, the southernmost province of Sweden; it was a rich flora, including many ferns, conifers, and other plants, but no flowering plants. More recently Professor Harris made a still larger collection during a long visit to eastern Greenland, in the ice-bound district of Scoresby Sound, where, under an extreme arctic climate, only a few stunted plants are able to exist. Nothing could be more striking than the contrast between the present floras of eastern Greenland and southern Sweden. The fossil plants of the Scoresby Sound district are of the same age as those from Scania, and demonstrate the former existence of a vegetation even richer and more varied than the contemporary vegetation of southern Sweden. A luxuriant and uniform vegetation occupied an area stretching from Central Germany to southern Sweden, and a thousand miles farther north beyond lat. 70° N. The fossils from these widely separated localities give no indication of any such change in the plant communities as we should expect and as we find when we contrast arctic and temperate floras in the modern world. It is possible that a clue to the solution of this, and similar problems, may be found in the theory of drifting continents, mentioned in Chapter II. There is another consideration relevant to the employment of fossil plants as tests of climate in former ages, which may be illustrated by this question: one asks whether it is justifiable to assume that the former presence of plants in Arctic regions, closely related to species now characteristic of warm temperate or even semi-tropical countries, necessarily implies that the climate was formerly the same, or nearly the same, as that of the regions where the modern descendants of the extinct species are now most at home? This question is usually answered in the affirmative, but a less orthodox view is perhaps worth stating. It is at least safe to say that the climate of Arctic regions was in earlier stages of geological history definitely warmer than it is to-day, not only in the Tertiary era, but, as the evidence of fossils proves, in Mesozoic and Palaeozoic periods. It is, however, by no means certain that the average temperature was as high as the present geographical distribution of the living relatives of extinct arctic species would seem to suggest. The reaction of a plant belonging to a genus which we know was in existence at least 80 or 100 million years ago is not necessarily the same as the reaction of its nearest living relatives to climatic conditions in earlier ages. The constitution of a plant—its sensibility to external influences—may well be a variable quality which suffers a change, a deterioration accompanying senility and loss of vigour. It does not follow that plants can be used as accurate thermometers of the ages. In the heyday of its youth a genus or a family may well have been much more tolerant of fluctuation in temperatures and unfavourable conditions than are their enfeebled descendants.

Not far from Cairo the surface of the Libyan desert is strewn with broken pieces of petrified stems of broad-leaved trees that had been embedded in sandy rocks, and by reason of their greater hardness withstood destruction by erosion. An even more impressive occurrence of large trunks of trees, their tissues transformed into jasper, chalcedony, and other flinty material, is the well-known petrified Triassic forest in Arizona which, like the Libyan trees, is reminiscent of days when 'rivers ran in dry places'.

Turning to naked-seeded trees we find their history goes back to much older periods. The oldest known conifers, for example, differ too widely from any living species to be assigned to the same genus: unfortunately none of the extinct plants have enabled us to see clearly the road leading to ancestors of a more primitive type. Conifers of many kinds, some closely related to living trees, others of a type long since extinct, were abundant in Jurassic forests, and several were in existence in much earlier ages. The genus *Araucaria*, now confined to South America,

Australia, New Guinea, New Caledonia and a few other Pacific islands, had a wide geographical range in the Jurassic period in the Northern Hemisphere, including England. Another much smaller group of naked-seeded plants—the Cycads—is of great interest from an evolutionary point of view. Cycads, or as some of them are called 'sago palms', are a small company almost confined to the tropics; the most abundant genus, *Cycas*, has a stem superficially resembling the columnar trunk of certain palms supporting a crown of long leaves bearing a double row of relatively long and narrow leaflets. The largest Cycads are African and have fronds several feet long: good specimens of this group may be seen in the tropical house in the Royal Botanic Gardens, Kew. Cycads are now sparsely scattered and are never conspicuous features of the vegetation; they occupy an isolated position in the plant kingdom. Many fossils from Jurassic and early Cretaceous rocks in temperate regions have been described as Cycads; the stems and leaves of these extinct plants agree closely with those of living genera, but their fertile shoots—the male and female organs—are, most of them, poles asunder from any existing reproductive structures. Two facts of special significance are: (i) the fossil Cycads, as they are called, were very much more widely distributed and much more numerous than existing Cycads; and (ii) the fossil forms differed considerably from the true Cycads of the present day in the plan of their reproductive organs, although in other respects they were very similar. On the other hand, some fossils exhibit a closer approach to living genera: the great majority of the fossils which it is customary to call Cycads are placed in a special group which has long been extinct. The prominent position and wide geographical range of these extinct plants in the latter part of the Mesozoic era is comparable to the position then held by extinct families of reptiles. The Tertiary era is often called the age of mammals and the age of flowering plants, so the latter part of the Mesozoic era is called the age of reptiles and the age of Cycads, that is, a group of plants superficially resembling living Cycads but distinguished by characters which have the greatest weight as evidence of affinity.

The large class of naked-seeded plants (Gymnosperms) includes, in addition to conifers and Cycads, another group which has now only one representative, the maidenhair tree, *Ginkgo biloba*. It is probably true to say that *Ginkgo* is the oldest tree in the world in the sense of length of years covered by its long line of predecessors of the same group. The maidenhair tree was formerly allocated to the family of conifers of which the best known member is the yew (*Taxus*), but it was subsequently removed to a group apart—the *Ginkgo* alliance—because of the discovery of certain peculiar and primitive characters. Several kinds of fossil plants from Jurassic and Cretaceous rocks were found to be more or less closely allied to *Ginkgo* and they afforded evidence of the world-wide distribution of the group in former ages. The maidenhair tree is often described as a 'living fossil', known only in cultivation; a tree which, had it not been carefully tended by man in China and Japan and venerated for certain supposed miraculous powers, would now be unknown in life. Some years ago a Chinese botanist stated that he had seen specimens growing naturally in forests in eastern China; be this as it may, China was no doubt its last home where it retained a hold on life and found a refuge as the sole survivor of a line all the other members of which had long been extinct. Leaves and a few seeds of many plants of the *Ginkgo* stock have been recorded from Tertiary, Cretaceous, Jurassic and Triassic rocks in nearly all parts of the world. A few leaves from still older rocks of Palaeozoic age are possibly, though not certainly, the foliage of related trees. The date of the first appearance of this once flourishing and far-flung group cannot be fixed with any precision. *Ginkgo* itself lived on the early Tertiary continent which extended from Polar regions to north-western Europe, and its forebears, with other closely allied types, were abundant in the forests of North America, Europe, South America, South Africa, and Australia, in the Triassic, Jurassic, and Cretaceous periods. By the end of the Tertiary era *Ginkgo* was the only genus left in Europe; it was doubtless driven eastwards by unfavourable climatic conditions during the Ice Age, and at length reached the Far East, where it remains as a symbol of persistence, almost a symbol of eternity.

The geological history of plants and animals, though it fails to satisfy the demand for the elusive missing links, demonstrates the rise and fall of many classes or groups, which after long periods of successful development, as measured by the production of new types and wanderings over broad spaces, passed completely, or in some instances almost completely, into oblivion. On the other hand, some of the common and little regarded plants on our British moorlands and hills are among the most ancient members of the vegetable kingdom, relics from the Palaeozoic era. Two living genera, *Selaginella*, a relatively small plant chiefly tropical in distribution and represented by a single species, the prickly club-moss, an ally of *Lycopodium* the ordinary club-moss but not related to true mosses, occasionally seen on the hills and bogs of Scotland, Wales, and the Lake District, and the genus *Equisetum* (horsetails), are among the oldest of all living plants. *Equisetum* has a wide distribution in Arctic lands, and in the temperate zone of both hemispheres. Fossils discovered among the debris of the Palaeozoic forests of the Coal Age show an amazing resemblance to those genera, both of which belong to groups which long ago included a large and varied assortment of trees very much larger and more complex in structure than any of the diminutive survivals. These and other races of plants illustrate the truth of the general statement based on the exploration of records of life scattered through the geological series, that the course of evolution has not followed a continuous upward trend from simple to less simple and increasingly more elaborate forms; it has also been retrogressive. Onward and upward, and occasionally downward, through long periods, is a fair description of the fluctuating and uneven process of development from age to age.

The oldest known plants in the world, apart from a few rather more ancient and fragmentary remains of seaweeds, have been described from Silurian beds in Australia: they are very similar in the plan of construction to plants previously found in early Devonian rocks in Scotland, Germany, and several other countries. Reference was made to these fossils in Chapter xv and little more need be said here. They are assigned to extinct groups of which it is probable no direct descendants survive. One of the Silurian plants, the largest of them, is known from stems about 2½ in. in diameter, occasionally forked and provided with crowded, long and narrow leaves up to 1½ in. in length, like blades of grass, and it is significant that at the base of some of the leaves were borne capsules containing spores. This genus (*Baragwanathia* after Mr Baragwanath, the discoverer of the fossils), so far as it is possible to describe it from the available material, reminds one of a fairly common club-moss (*Lycopodium selago*) familiar to anyone who has noticed the low-lying vegetation on moorland and mountains, a species of unusually wide range from Arctic regions, deep into the Southern Hemisphere. It is not improbable that the Silurian genus may be a distant connexion of our club-mosses. Another plant, *Zosterophyllum*, which lived in the latter part of the Silurian period, and in the early part of the Devonian period, was not more than about 6 or 8 in. high; its short stem bore a tuft of leafless narrow branches, to some of which, near the tips, were attached kidney-shaped spore-cases, containing spores suitable for dispersal by wind. *Zosterophyllum*, with others of the oldest known land plants, had neither roots nor leaves; it cannot be closely compared with any genus in the present age. Though in some respects a primitive type, its organization is not as simple as might be expected; there can be no doubt it, and its associates, were descendants of still more ancient ancestors of which nothing is known. We have no knowledge of the vegetation which grew on Ordovician and Cambrian lands. No trees have been found in the older Palaeozoic plant beds: the probability is that several of the plants were only partially terrestrial, and were able to grow in shallow water. Possibly they were pioneers of a land flora, and in an earlier age were represented by simpler types living in the sea, but this is pure speculation.

There is a possibility that the genera *Selaginella* and *Equisetum*, described as two of the most ancient living plants, may have to give place to a less familiar plant, *Psilotum*, one of two tropical and sub-tropical genera formerly classed with the club-mosses, and now placed in a group of their own. *Psilotum*, grown under glass in botanical gardens, ranges from Florida to New Zealand and Japan; it is from $\frac{1}{2}$ to 2 ft. in height, its slender, angular stems are repeatedly forked and bear minute, hardly visible scale-like leaves, and small orange-coloured spore-capsules; it has no roots. Many of the oldest known fossil plants from Devonian and Silurian rocks have been compared with *Psilotum*: there are important differences, and there are also resemblances which may mean distant relationship. If *Psilotum*, which now occupies an isolated and detached position with no near relatives except its one much rarer companion (*Tmesipteris* of New Zealand), is entitled to be considered a possible link with a group of Devonian and Silurian plants separated from the present day by 400 million years, it is surprising that no fossils are known from rocks later than Devonian which afford evidence of affinity to this faint reflection of the oldest known members of a land flora. The oldest plants of the land so far discovered are samples of a vegetation at a locality near the sea-coast; the remains were found in a marine sediment. There must have been other plants in the hinterland, but of them we are completely ignorant.

Two additional examples of plants that have a special claim to be called survivals are worthy of mention. Many people are familiar, at least by sight, with the small green plants called by botanists *liverworts*, allies of mosses: a common form, abundant on garden paths and the damp surface of walls, is an inch or two in length, consisting of a repeatedly forked flat body; other forms are erect, and the body consists of a thread-like stem bearing very delicate, minute leaves. Liverworts are known to have lived in the forests of the Coal Age, and good examples of both flat, leafless species and leafy species have been described by Professor Walton of Glasgow; they differ hardly at all in general appearance and structure from some existing genera. A brief account of a Rhaetic liverwort was given in Chapter VI.

Little has been said in earlier chapters of the past history of the large class Algae, that is, seaweeds as well as fresh-water plants and some that live on land. Seaweeds, comparable with living genera, lived in Cambrian seas and in seas of all ages. Perhaps the most remarkable example of an organism which has existed with no recognizable modification in all probability from the Ordovician period and certainly from the Carboniferous period until the present day is a very small alga called *Botryococcus braunii*. This simple plant deserves special treatment as a type that continued its course through the ages, retaining apparently its vigour oblivious to the passage of aeons of time, an amazing emblem of unalterability and endurance. *Botryococcus braunii* has been found in temperate and tropical countries practically all the world over; it usually lives in fresh-water lakes and ditches, and sometimes in the sea. With many other Algae it occurs in the floating, microscopic vegetation on

Windermere and many other lakes; its colour varies from green to orange. An interesting peculiarity is its habit of exuding drops of oil; it is an oil-producing plant. Another quality is its ability to live through periods of desiccation, and it is occasionally found on land. *Botryococcus* consists of minute pear-shaped cells more or less regularly disposed in radially arranged rows embedded in a transparent framework secreted by the living cells. These groups of cells, with this encasement, are spoken of as colonies and are often approximately spherical. Each cell is enclosed in a waterproof thimble-shaped cup, and this is very resistant to decay. In many places masses of colonies consisting of the framework have been found as pieces or layers of sticky black jelly of a rubber-like texture, and remains of the alga are recorded from beds of peat, the mud of lakes and on the shores of lakes and lagoons, as in the district of Coorong, near Adelaide in South Australia. The plant multiplies in the simplest way by division of each cell into two: this is the only method of reproduction so far discovered. Now let us see what evidence there is of the antiquity claimed for *Botryococcus*: fossil Algae, closely allied to the living species, are known from Ordovician and other rocks of the earlier part of the Palaeozoic era, but much more attention has been paid to Algae of this type from Carboniferous rocks. Nearly a century ago a sort of cannel coal called Boghead and Torbanite, from the places Boghead and Torbane Hill in the Midland Valley of Scotland, not far from Edinburgh, was used as a source of gas and paraffin: some years later, this material was examined microscopically and it was seen to consist largely of small yellow bodies which, on high magnification, stood out conspicuously as yellow or sherry-coloured spots in a dark ground. In 1889, Sir T. Edgeworth David, Professor of Geology at Sydney, noticed precisely similar yellow bodies in some oil shales in New South Wales, known as kerosene shale, and he was the first to suggest the possibility of their algal nature. But, despite the fact that the yellow bodies in the Australian kerosene shale and the Boghead of Scotland showed what appeared to be a cellular structure resembling that of some Algae, there was general scepticism about the botanical nature of the bodies, chiefly because it seemed very improbable that such delicate structures could be preserved. A long controversy was kept up for many years—are the yellow bodies plants or are they resinous deposits which have assumed a structure simulating plant cells? In 1914, a Russian geologist compared the yellow bodies with the alga *Botryococcus*, and expressed the opinion that the yellow bodies which had meanwhile been described as extinct Algae and given special names, were in reality fossil colonies of the living alga. Dr Kathleen Blackburn and Dr Temperley, in a paper published in the *Transactions of the Royal Society of Edinburgh* in 1936, made out a convincing case in support of the practical identity of the Carboniferous fossils, and the living *Botryococcus braunii*. It should be added that this alga has also been discovered in Rhaetic beds as well as in rocks of later periods. *Botryococcus* may be described as a plant that is heir of all the ages, a product of evolution that came into existence at an early stage in the development of the vegetable kingdom, and followed the even tenor of its way, changeless in a world of ceaseless change. It may have been an early stage in the upward progress to higher and higher forms of plants, but be this as it may, the living *Botryococcus* speaks to us as an echo from a world in its early youth, and bridges a gap measured by some hundreds of millions of years.

Enough has been said to show that the plant records of the rocks offer an attractive line of enquiry which has already been fruitful in results; it is a branch of botanical and geological science well worthy of study by naturalists whose interest in plants is not limited to the age in which we live. The best that can be offered is a very imperfect picture of the past, but none the less a picture of surpassing interest despite, or perhaps because of, its imperfections. Man, whose days are 'as it were a span long', is privileged to read the tattered pages of Nature's story-book. As Lucian said nearly 1800 years ago:

> The world is fleeting; all things pass away;
> Or is it we that pass and they that stay?

The manner of unfolding of life remains a mystery; the theory of Natural Selection, as formulated by Darwin and by Wallace, no longer finds favour with many biologists; but, whether or not it will be reaffirmed or abandoned, the central conception of evolution remains as an article of belief in the naturalist's creed. The memorable and familiar words with which Darwin concluded his greatest work have not lost their force; their nobility and simplicity are a fitting profession of faith by one of the greatest thinkers and noblest characters of the Victorian era:

Thus, from the war of Nature, from famine and death, the most exalted object which we are capable of conceiving, namely the production of the higher animals, directly follows. There is grandeur in this view of life, with its several powers, having been breathed into a few forms or into one; and that, whilst this planet has gone cycling on according to the fixed law of gravity, from so simple a beginning endless forms most beautiful and most wonderful, have been and are being evolved.

INDEX

For EU product safety concerns, contact us at Calle de José Abascal, 56–1°,
28003 Madrid, Spain or eugpsr@cambridge.org.